DATE DUE

~~JA 1 '95~~			
~~NO 3 '95~~			

DEMCO 38-296

NEW HORIZONS

**ALSO BY
P. D. OUSPENSKY**

NEW HORIZONS

Explorations in Science

by
P. D. Ouspensky

With an Introduction by
Colin Wilson

ISBN 0-936385-21-9
Library of Congress Catalog Card No. 90-082356
10 9 8 7 6 5 4 3 2 1
Manufactured in the United States

CONTENTS

EDITOR'S NOTE

Recent years have seen the publication of a variety of books on the new physics, the wonders of the sub-atomic world, incredible changes in theories of cosmology and the origins and nature of the universe, and the seemingly "mystical" findings of physicists at the very boundaries of human knowledge. Speculation has become an ordinary occupation for writers on science and scientific theories. However, even in the most esoteric of these writings, the name of Peter Demianovitch Ouspensky is remarkably absent.

It has been almost 60 years since the appearance of the essays in this book within the confines of the massive compendium Ouspensky called A New Model of the Universe. Mathematician, philosopher, theorist and writer, Ouspensky's ventures into theoretical science found little appreciation in their own day. Like many before him, his speculations seemed "fantastic" to his contemporaries. Today's readers are much better prepared to digest the sweeping range of his thoughts, the daring of his intellectual conjectures, and the breathtaking vista he creates. We recognize in Ouspensky's writings the patterns of a world that, in our own time, is opening to investigation. Therefore, it seems a propitious time to look once again at these writings of one of the first of modern thinkers to see the connections existing between the observed world of investigative science and the unseen world of mystical reality.

In selecting the essays included in this book, we have tried to focus attention on the range of Ouspensky's writings within scientific speculation. Originally written between 1905 and 1912, they were continually revised until 1929, to reflect the changes in his thinking. It is astonishing that Ouspensky could have foreseen with such clarity the direction of contemporary speculation at such an early date, and these essays, therefore, could well be described as classics in their field. For anyone interested in the speculative reconstruction of the known and unknown universe, we believe these essays will be of compelling interest, throwing an entirely new light on current thought.

One of the most consistent impressions in reading these pieces is the insistence on what Ouspensky calls "esoteric thinking". In each essay, the author has woven into the fabric of scientific speculation the threads of esotericism. For instance, we find him returning again and again to ideas of higher knowledge, altered realities and potential human functions, to name a few. Although this element is very strongly expressed, it is also very much at the center of our reasons for having chosen to republish this book at the present time.

Colin Wilson, the prolific English author of such works as *The Outsider*, *The Occult*, and *The War Against Sleep*, has graciously added an introduction to Ouspensky and his work. These essays are reprinted from the Kegan Paul, Trench, Trubner & Company edition of 1934, published in London.

INTRODUCTION

by Colin Wilson

In the spring of 1918, a young Russian named Nicholas Bessarabov knocked on the door of the American writer and architect Claude Bragdon, who lived in Rochester, N.Y., and told him that he had just encountered one of the most exciting books he had ever read. It was called *Tertium Organum*, and it was by a totally unknown Russian philosopher named Peter Ouspensky. Bragdon, who could read Russian, found it equally exciting, and agreed to help Bessarabov translate it. Bragdon's enthusiasm was so great that he decided to publish the book himself. To his delight and astonishment, it was an instant success, selling 7,000 copies in the course of one year.

Its author, who fled from the Russian Revolution, and was living in Constantinople, was equally delighted to receive some unexpected royalties from America, and asked Bragdon if he could help him to immigrate to England. A few weeks later, with the aid of another Ouspensky admirer—Lady Rothermere, the wife of the newspaper magnate—Bragdon was able to send Ouspensky the money to travel to London.

He arrived in August 1921, and discovered that he was already something of a celebrity. *Tertium Organum* had gone into a second edition, and Lady Rothermere had given away copies to everyone she thought might be interested. Ouspensky was introduced to A. R. Orage, one of the most influential editors of the day, and within two months was lecturing to a group of admirers at a studio in St. John's Wood. Within a short time he had become one of the most powerful intellectual influences in postwar London.

Who was this messiah, and how had he acquired that air of immense certainty with which he dominated his audiences—which included some of London's most prominent intellectuals? Pyotr Demianovitch Ouspensky had been born in Moscow on March 5, 1878, the son of an officer in the Survey Service and of a talented artist. At the age of six, the child began to experience that peculiar "I have been here before" feeling, and it became an obsession. Expelled from school at the age of

1

fifteen for a practical joke, Ouspensky soon discovered Nietzsche, and concluded that his "I have been here before feeling" was what Nietzsche meant by "Eternal Recurrence", and that it proves that human beings live their lives over and over again. At the age of eighteen, the death of his mother enables him to begin his "search for truth", and he began a series of travels around Russia and Europe—and later India—to try to find teachers who could initiate him into the secret of life. He wrote a rather gloomy novel, *The Strange Life of Ivan Osokin*, about Eternal Recurrence, and then, at the age of thirty three, began his own highly personal attempt to summarize the mysteries of human existence, *Tertium Organum*. (The title refers to the fact that Aristotle had written an *Organum* and Bacon a *Novum Organum*; Ouspensky's was the third.) And so great was his momentum, and his sense of new and important discoveries, that he immediately went on to write an even longer and more complex work, *A New Model of the Universe*. And it was only when he had almost completed this new book, in the spring of 1915, that he met the man who was to finally hand him the key to the "Hidden Knowledge" that he had been seeking for the past twenty years, George Ivanovitch Gurdjieff. And in due course he produced his third major work, the book in which he describes his years as a pupil of Gurdjieff, *In Search of the Miraculous*, finally published after his death in 1948.

Those who first made the acquaintance of Ouspensky's writings through this third book are at a considerable disadvantage. For, as everyone who knows Gurdjieff's work will be aware, Gurdjieff's teaching is of such remarkable and striking originality that it can make an impression of revelation. He asserts that human beings are so mechanical that they possess virtually no free will, and that everyday consciousness is, quite literally, a form of sleep. So to talk about truth or salvation or mystical insight is a waste of time; the first thing we have to do is to *realise we are asleep*. Then we can begin...

So anyone who comes to Ouspensky's work after reading *In Search of the Miraculous* will probably feel, as I did myself, that *Tertium Organum* and *A New Model of the Universe* are merely "prentice works" that hardly deserve a second glance. And this, in fact, turns out to be a mistake of the first order. For the truth is that Ouspensky can no more be dismissed as a disciple of Gurdjieff than Coleridge could be dismissed as a disciple of Wordsworth, or Schiller of Goethe. He is one of the most remarkable and original thinkers of the twentieth century, and even if he had died in 1915, immediately after finishing *A New Model of the Universe*, he would still rank as one of the most outstanding philosophers to come out of Russia.

I have to admit that I came to this realisation at an absurdly late stage. It was in the autumn of 1987, and I was preparing to leave my home in Cornwall to pay a visit to America, during the course of which I was scheduled to give some lectures and interviews. On the morning I was due to leave, I noticed my copy of *A New Model of the Universe* lying beside the bed—I had taken it off the shelf to duplicate some pages about Eternal Recurrence for a friend. I opened it casually to the chapter called "Experimental Mysticism", and observed, from the markings in the margin, that I must have read it before. And since I had no recollection of having done so, I proceeded to read it again. Within five minutes, I was aware that this was one of the most exciting and original pieces about mysticism that I had ever read. When I had finished it, I duplicated the chapter. I re-read it on the plane to New York, and then talked about it non-stop for the next three weeks.

The odd thing, of course, is that I had read it before without recognising its seminal importance—even though I had marked some of its most interesting passages. The answer, obviously, is that I was not ready for it, and missed half its significance. Ouspensky makes the same point in an important passage in *Tertium Organum*—Chapter 14—where he says: "It seems to us that we see something and understand something. But in reality all that proceeds around us we sense only very confusedly, just as a snail senses confusedly the sunlight, the darkness and the rain." And he goes on to tell how he and a friend were walking across the bridge over the Neva, looking at the Peter and Paul Fortress, when the friend remarked that there were factory chimneys inside the prison. "On his saying this, I too sensed *the difference* between the chimneys and the prison walls *with unusual clearness* and like an electric shock."

What he is saying is of tremendous importance, and yet hardly expressible in words. But the phrase "like an electric shock" catches something of its vividness. We can find the same thing elsewhere in the book, chapter twelve, in a description of a " mystical" experience when he was crossing the sea of Marmora on a steamer in heavy seas. " I watched this play of waves with the ship, and felt them draw me to themselves. It was not the desire to jump down which one feels in mountains but something infinitely more subtle. The waves were drawing my soul to themselves. And suddenly I felt it went to them. It lasted an instant, perhaps less than an instant, but I entered into the waves and with them rushed with a howl at the ship. And in that instant *I became all*. The waves—they were myself: the far violet mountains, the wind, the clouds hurrying from the north, the great steamship heeling and rushing irresistibly forward—all were myself...It was an instant of unusual freedom, joy and expansion. A second later, the spell of charm disappeared."

These passages also emphasize a point made by Bragdon in his introduction: that "Ouspensky's clearness of thought is mirrored in a corresponding clarity of expression." Ouspensky writes with a precision that combines the poet and the scientist. To read *Tertium Organum* and *A New Model of the Universe* is to share his experience in the way that we share the experience of a great travel writer or autobiographer. This is not abstract, plodding stuff; it has the authentic taste of real life.

Now Gurdjieff is obviously one of the greatest thinkers (or prophets, or whatever you prefer to call it) of the twentieth century—I personally would place him far ahead of all the rest. And coming upon his ideas was, for Ouspensky, one of the greatest pieces of luck he ever encountered. Yet the fact remains that Ouspensky possessed his own unique greatness, and anyone who has never read *Tertium Organum* or *A New Model of the Universe* or *The Strange Life of Ivan Osokin* has missed a remarkable and immensely satisfying experience. For, as odd as it sounds, Ouspensky was not merely a highly individual thinker; he was also a considerable artist.

I say "as odd as it sounds" because Ouspensky did not strike people as the artistic type—anything but. One of his first English followers, Rowland Kenney, was obviously rather repelled by his first encounter. "His nose made one think of a bird's strong beak; indeed, when sitting in reflection or repose he hunched himself together and looked like a dejected bird huddling up in a rainstorm. He was obviously a man of a dominant if not domineering type of character, with determination—or obstinacy—written over his every feature." Another of these early acquaintances, Paul Selver, found him "quite monumentally boorish.

He was one of those exasperating Russians who refuse to credit any other Slavic nation with artistic ability..." Ouspensky was a strong character who felt that he had stumbled upon some important truths that were virtually unsuspected by the rest of humanity, and he was disinclined to waste time or energy talking "on equal terms" with complacent intellectuals. When one silly lady in his audience asked if the Buddha was on the seventh level of consciousness, he replied simply "I don't know", and in the ensuing silence left her to blush for her own idiocy.

Soon after Ouspensky's first meeting with Gurdjieff, Ouspensky asked whether anything useful could be achieved by studying "occult" or mystical literature. He had in mind such systems as the Tarot. Gurdjieff replied that a great deal could be learned from reading. "Take yourself: you might already know a great deal if you *knew how to read*. I mean that if you *understood* everything you have read in your life, you would already know what you are looking for now. If you understood everything you have written in your own book...I should come and bow down to you and beg you to teach me. *But you do not understand* either what you read or what you write."

A New Model of the Universe, from which all the selections in *New Horizons* are taken, is an interesting illustration of what Gurdjieff meant; yet it also demonstrates that Gurdjieff failed to do justice to Ouspensky as an original thinker. It was written, in fact, after Ouspensky had met Gurdjieff, and this in itself is an indication of Ouspensky's independence. The meeting with Gurdjieff was a watershed in his life, and one would expect the book to be full of hints of the new mental horizons that had been opened up for him. In fact, he had so far developed his own original ideas and insights that he had no need to speak about his latest discoveries; the ideas he had developed since *Tertium Organum* were enough to fill this very substantial volume.

The "Author's Introduction" to *New Horizons* is a marvelous example of Ouspensky's ability to seize and hold the reader's attention. When, in the late 1960's, I began to write *The Occult*, I chose it as my starting point: Ouspensky sitting in a Moscow newspaper office, faced with a huge pile of foreign newspapers, from which he has to concoct an article on the Hague Conference. Groaning with boredom, he opens his desk drawer, in which he has books with such titles as *The Occult World*,

Atlantis and Lemuria and *The Temple of Satan*, and plunges immediately into a world that seems to him infinitely more exciting and significant than that of politicians and their platitudes and lies...

But this "Author's Introduction" also makes us aware of a certain naivete in the young Ouspensky. Few of the books he mentions, Eliphas Levi's *Ritual Magic* is an example—would be taken seriously by one of today's "alternative thinkers". Ouspensky is obviously as romantic—and gullible—as the young W.B. Yeats of the 1890's. And his basic idea— that somewhere in the world there are secret societies that possess "esoteric wisdom"—sounds like a leftover from the same period. Yet it is also clear that, even in these early days, Ouspensky sensed that something new and strange was happening. "..it seemed to me... that there is evidence, if not of a new race, at least of some new category of men for whom there exist different values than for other people." In the 1930's, in his novel *Star Begotten*, H.G. Wells would invent the term "Martians" to describe these "new men", and twenty years later, I christened them "Outsiders".

But by the time he returned from his "search for secret wisdom" in the East, Ouspensky had already developed his own highly original philosophical system. The starting point of that system is that, although western science provides the basic model for *all* knowledge, western scientists have deliberately blinkered themselves to the different *kinds* of knowledge to be found in mysticism, religion and magic. Ouspensky *was* a scientist, and remained so all his life; but he wanted to apply science to the unknown regions of human consciousness. Half a century later, Aldous Huxley embarked on the same quest when he decided to take mescaline, and was staggered by the revelation of "unknown modes of being". Ouspensky knew about peyote, but his starting point was the nitrous oxide experience (already explored by William James), hypnosis and the study of dreams. No chapter gives a better measure of his originality than the one on the study of dreams and hypnosis. He was deeply interested in hypnagogic states—those borderland states be-tween sleep and waking—set out to cultivate the ability to observe and remember them. His reports on what he calls "half-dream states" demonstrates how far it is possible for a lone individual to achieve his own insights into these "unknown modes of being". Readers who are fascinated by this chapter are advised to go on to the classic study

Hypnagogia by Professor Andreas Mavromatis (1987), which describes the remarkable results of some group experiments in "half-dream states", and which repeatedly acknowledges its debt to Ouspensky's brilliant pioneering work.

I must admit that I was at first puzzled why the editor had chosen the chapters *The Fourth Dimension* and *A New Model of the Universe* for inclusion. Yet as soon a I re-read these chapters, I understood precisely why. They are among the best examples of Ouspensky's ability to create a sense of sheer intellectual excitement. Ever since Einstein, the subject of the fourth dimension has been regarded as rather out of date, for most of us are under the impression that Einstein "proved" that the fourth dimension is time. This is typical of the kind of short- sightedness that creates intellectual fashions. We need to remember that in the 1870's, the concept of the fourth dimension created enormous intellectual excitement, and seemed to be the answer to many puzzling questions about the universe. Professor Johann C. F. Zöllner, of Leipzig University, became convinced that this mathematical concept was the only way of explaining some of the mysteries of psychical research, and he concluded that "the dead" must live in a world that has one dimension more than our own. If that is so, he reasoned, then a "spirit" ought to be able to tie a knot in a piece of cord whose two ends were joined. At a seance in 1877, in the presence of a medium called Slade, a knot *was* tied in a circular piece of cord, apparently verifying Zöllner's theory. Sceptics suggested that the medium had simply switched cords, which may be so. But Ouspensky was convinced that the fourth dimension was the "key to the enigmas of the world" (the sub-title of *Tertium Organum*), and most of *Tertium Organum* is devoted to a discussion of the concept. And in Chapter Eleven of that book, he quotes a fascinating extract from Johan Van Manen in which Van Manen describes how, one night, as he lay in bed trying to visualise the fourth dimension, "I saw plainly before me first a four-dimensional globe and afterwards a four dimensional cube..." Unless Van Manem was deceiving himself (and he claimed to be able to recall the four dimensional globe with ease, and the cube with more difficulty), then our minds *are* capable of grasping the idea of extra dimensions, and we can begin to see why Ouspensky felt that this was the basic key to the understanding of the universe. The chapter *A New Model of the Universe* attempts to develop the scientific implications of this theory, which is why no selection from Ouspensky would be complete without it.

But the most important chapter in the book—and probably in all Ouspensky's work—is the one called *Experimental Mysticism*. Here Ouspensky describes certain experiments he undertook in 1910 and 1911. He is deliberately vague about his methods, but from a comment about narcotics he made to Gurdjieff ("I have myself carried out a number of experiments in this direction") it is almost certain that he made use of some drug, possibly nitrous oxide ("dental gas"). And—as the chapter makes clear—this experience gave him a completely new insight into the fourth dimension and the problem of life after death.

One modern experimenter, R. H. Ward, remarked in his book *A Drug-Taker's Notes*: "On this occasion it seemed to me that I passed, after the first few inhalations of the gas, directly into a state of con-sciousness already far more complex than the fullest degree of ordinary consciousness." And in saying this, he has made a point of crucial importance for the study of human potential. It seems natural for us to accept our present state of consciousness as fundamentally "true" and inevitable. And, what is worse, we accept the "duller" and narrower states of consciousness as somehow truer than the wider and more exciting ones, in the same way that we accept sobriety as "truer" than drunkenness. The insight that came to Ward is that our everyday consciousness is essentially *incomplete*. It follows that if we actually *knew* this, we would readjust our response to existence, particularly in the matter of allowing ourselves to become pessimistic and "discouraged".

Ouspensky's first observation in this new state of consciousness was that it was impossible to say anything about it—not for the usual reason described by mystics, that language is inadequate, but because "there is nothing separate, that is, nothing that can be described or named *separately*. In order to describe the first impressions...it is necessary to describe *all* at once. The new world with which one comes into contact has no sides, so that it is impossible to describe first one side and then the other. All of it is visible at once at every point." In other words, he saw that *everything is connected*. William James had made much the same comment about his own "mystical experiences": "What happened each time was that I seemed...to be reminded of a past experience; and this reminiscence, ere I could conceive or name it distinctly, developed into something further that belonged with it, this into something further still, and so on, until the process faded out, leaving me amazed at the sudden vision of increasing ranges of distant facts of which I could give no articulate account." In other words, the first memory reminded him of something else, and that reminded him of something else, and that reminded him of something else...so that it was rather like a flash of lightning zig-zagging out towards "distant horizons." And it is interesting to note that James speaks of "increasing ranges of distant *facts*", ordinary everyday facts, not some visionary or mystical insight.

We have all had some similar experience in moments of great excitement or enthusiasm—even that feeling on spring mornings, or setting out on holiday, when everything you look at seems to bring back other times and other places. Ouspensky said that as he looked at an ashtray, his mind could grasp so many *implications*—the mines where its metal was obtained, the men who had transported it and made it, and so on—that he wrote on a piece of paper "One could go mad from one ashtray."

So, in a sense, Ouspensky already knew all that Gurdjieff had to teach him; he had already glimpsed the immense possibilities of human consciousness, and seen that the problem lies in our "mechanicalness"—he says that when he came back to the "normal world" after his mystical experiences, it was like returning to a "wooden world", "as if it were an enormous wooden machine with creaking wooden wheels, wooden thoughts, wooden moods, wooden sensations..." Then why did he need Gurdjieff to teach him? The answer is already implied in what has just been said: he failed to grasp the *implications* of his insights, their connections with other insights. Gurdjieff was right when he said that if Ouspensky had been able to understand his own writings, he would have been a sage.

Now from Ouspensky's point of view, that was unfortunate. From ours, it is the reverse. Ouspensky is such a beautifully clear writer, and so transparently honest, that we can enjoy his work purely as a literary experience, and absorb more of its implications with every reading. It is sad that, by comparison, Gurdjieff was such an atrocious writer.

In 1935, Rom Landau published a book called *God is my Adventure*, a study in various modern "prophets", which became an instant bestseller, and which is still a fascinating and delightfully readable book.

What will baffle modern readers is that he devotes two chapters to Ouspensky and Gurdjieff (in that order), and appears to make little or no connection between them (in spite of the fact that Ouspensky acknowledged his debt to Gurdjieff.) The chapter on Ouspensky is called "War Against Sleep" (a title I stole for my own book on Gurdjieff), and expounds Gurdjieff's ideas as if they were Ouspensky's own. The chapter on Gurdjieff (called "Harmonious Development of Man") is thoroughly hostile, and implies that Gurdjieff is a mercenary charlatan. Landau's view prevailed for the next fifteen years until, after the deaths of both Gurdjieff and Ouspensky, and publication of Ouspensky's *In Search of the Miraculous*, the true stature of Gurdjieff began to be recognised. (I played my own small part in popularising Gurdjieff's teaching with a section on him in *The Outsider* in 1956.) But the pendulum swung too far in the opposite direction. Instead of being regarded as a highly original thinker in his own right, Ouspensky was suddenly reduced to the status of a peddlar of secondhand ideas. As readers of *New Horizons* will discover, nothing could be more absurdly unfair. If those same readers can be persuaded to go on to *Tertium Organum* and *A New Model of the Universe*, I think I can promise them one of those profound intellectual experiences that will continue to reverberate for a lifetime.

AUTHOR'S INTRODUCTION

THERE exist moments in life, separated by long intervals of time, but linked together by their inner content and by a certain singular sensation peculiar to them. Several such moments always recur to my mind together, and I feel then that it is these that have determined the chief trend of my life.

The year 1890 or 1891. An evening preparation class in the Second Moscow " Gymnasium ".[1] A large class-room lit by kerosene lamps with large shades. Yellow cupboards along the walls. Boarders in holland blouses, stained with ink, are bending over their desks. Some are immersed in their lessons, some are reading under their desks a forbidden novel by Dumas or Gaboriau, some are whispering to their neighbours. But outwardly they all look alike. At the master's desk sits the master on duty, a tall lanky German, " Giant Stride ", in his uniform—a blue tailcoat with gold buttons. Through an open door another such preparation class is seen in the adjoining class-room.

I am a schoolboy in the second or third " class ". But instead of Zeifert's Latin grammar, entirely consisting of exceptions which I sometimes see in my dreams to this day, or Evtushevsky's " Problems ", with the peasant who went to town to sell hay, and the cistern to which three pipes lead, I have before me Malinin and Bourenin's " Physics ". I have borrowed this book from one of the older boys and am reading it greedily and enthusiastically, overcome now by rapture, now by terror, at the mysteries which are opening before me. All round me walls are crumbling, and horizons infinitely remote and incredibly beautiful stand revealed. It is as though threads, previously unknown and unsuspected, begin to reach out and bind things together. For the first time in my life my world emerges from chaos. Everything becomes connected, forming an orderly and harmonious whole. I understand, I link together, a series of phenomena which were disconnected and appeared to have nothing in common.

[1] " Gymnasiums " were government " classical " schools containing eight classes, i.e., forms, for boys from ten to eighteen.

11

But what am I reading?

I am reading the chapter on levers. And all at once a multitude of simple things which I knew as independent and having nothing in common become connected and united into a great whole. A stick pushed under a stone, a penknife, a shovel, a see-saw, all these things are one and the same, they are all "levers". In this idea there is something both terrifying and alluring. How is it that I did not know it? Why has nobody spoken to me about it? Why am I made to learn a thousand useless things and am not told about "this"? All that I am discovering is so wonderful and so miraculous that I become more and more enraptured, and am gripped by a certain presentiment of further revelations awaiting me. It is as though I already feel the *unity of all* and am overcome with awe at the sensation.

I can no longer keep to myself all the emotions which thrill me. I want to try to share them with my neighbour at the desk, a great friend of mine with whom I often have breathless talks. In a whisper I begin to tell him of my discoveries. But I feel that my words do not convey anything to him and that I cannot express what I feel. My friend listens to me absent-mindedly, evidently not hearing half of what I say. I see this and feel hurt and want to stop talking to him. But the tall German at the master's desk has already noticed that we are "talking" and that I am showing something to my friend under the desk. He hurries over to us and the next moment my beloved "Physics" is in his stupid and unsympathetic hands.

"Who gave you this book? You can understand nothing in it anyway. And I am sure you have not prepared your lessons."

My "Physics" is on the master's desk.

I hear round me ironical whispers and comments that Ouspensky reads physics. But I don't care. I shall have the "Physics" again to-morrow; and the tall German is all made up of large and small levers!

Year after year passes by.

It is the year 1906 or 1907. The editorial office of the Moscow daily paper *The Morning*. I have just received the foreign papers, and I have to write an article on the forthcoming Hague Conference. French, German, English, Italian papers. Phrases, phrases, sympathetic, critical, ironical, blatant, pompous, lying and, worst of all, utterly automatic, phrases which have been used a thousand times and will be used again on entirely different, perhaps contradictory, occasions. I have to make a survey of all these words and opinions, pretending to take them seriously, and then, just as seriously, to write something on my own account. But what can I say? It is all so tedious. Diplomats and all kinds of statesmen will gather together and talk, papers will approve or disapprove, sympathise or not sympathise. Then everything will be as it was, or even worse.

It is still early, I say to myself: perhaps something will come into my head later.

Pushing aside the papers I open a drawer in my desk. The whole desk is crammed with books with strange titles, *The Occult World*, *Life after Death*, *Atlantis and Lemuria*, *Dogme et Rituel de la Haute Magie*, *Le Temple de Satan*, *The Sincere Narrations of a Pilgrim*, and the like. These books and I have been inseparable for a whole month, and the world of Hague Conferences and leading articles becomes more and more vague, foreign and unreal to me.

I open one of the books at random, feeling that my article will not be written to-day. Well, it can go to the devil! Humanity will lose nothing if there is one article the less on the Hague Conference.

All these talks about a universal peace are only Maniloff's dreams about building a bridge across the pond.[1] Nothing can ever come out of it, first of all because the people who start conferences and those who are going to debate on peace will sooner or later start a war. Wars do not begin by themselves, neither do "peoples" begin them, however much they are accused of it. It is just those men with their good intentions who are the obstacle to peace. But is it possible to expect that they will ever understand this? Has anybody ever understood his own worthlessness?

A great many wicked thoughts occur to me about the Hague Conference, but I realise that none of them are printable. The idea of the Hague Conference comes from very high sources; therefore if one is to write about it at all, one must write sympathetically, especially as even those of our papers which are generally the most suspicious and critical of all that comes from the government disapprove only of the attitude of Germany to the conference. The editor would therefore never pass what I might write if I say all that I think. And if by some miracle he were to pass it, it would never be read by anybody. The paper would be seized in the streets by the police, and both the editor and I would have to make a very long journey. This prospect does not appeal to me in the least. What is the use of attempting to expose lies when people like them and live in them? It is their own affair. But I am tired of lying. There are enough lies without mine.

---

[1] Maniloff, a sentimental landowner in Gogol's *Dead Souls*.

But here, in these books, there is a strange flavour of truth. I feel it particularly strongly now, because for so long I have held myself in, have kept myself within artificial " materialistic " bounds, have denied myself all dreams about things that could not be held within these bounds. I had been living in a desiccated and sterilised world, with an infinite number of taboos imposed on my thought. And suddenly these strange books broke down all the walls round me, and made me think and dream about things of which for a long time I had feared to think and dream. Suddenly I began to find a strange meaning in old fairy tales ; woods, rivers, mountains, became living beings ; mysterious life filled the night ; with new interests and new expectations I began to dream again of distant travels ; and I remembered many extraordinary things that I had heard about old monasteries. Ideas and feelings which had long since ceased to interest me suddenly began to assume significance and interest. A deep meaning and many subtle allegories appeared in what only yesterday seemed to be naïve popular fantasy or crude superstition. And the greatest mystery and the greatest miracle was that the thought became possible that death may not exist, that those who have gone may not have vanished altogether, but exist somewhere and somehow, and that perhaps I may see them again. I have become so accustomed to think " scientifically " that I am afraid even to imagine that there may be something else beyond the outer covering of life. I feel like a man condemned to death, whose companions have been hanged and who has already become reconciled to the thought that the same fate awaits him ; and suddenly he hears that his companions are alive, that they have escaped and that there is hope also for him. And he fears to believe this, because it would be so terrible if it proved to be false, and nothing would remain but prison and the expectation of execution.

Yes, I know that all these books about " life after death " are very naïve. But they lead somewhere ; there is something behind them, something I had approached before ; but it frightened me then, and I fled from it to the bare and arid desert of " materialism ".

The " Fourth Dimension " !

This is the reality which I dimly felt long ago, but which escaped me then. Now I see my way ; I see my work, and I see where it may lead.

The Hague Conference, the newspapers, it is all so far from me. Why is it that people do not understand that they are only shadows, only silhouettes, of themselves, and that the whole of life is only a shadow, only a silhouette, of some other life ?

Years go by.

Books, books, books. I read, I find, I lose, I find again, again I lose. At last a certain whole becomes formed in my mind. I see the unbroken line of thought and knowledge which passes from century to century, from age to age, from country to country, from one race to another; a line deeply hidden beneath layers of religions and philosophies which are, in fact, only distortions and perversions of the ideas belonging to the line. I see an extensive literature full of significance which was quite unknown to me until recently, but which, as now becomes quite clear to me, feeds the philosophy we know, although it is scarcely mentioned in the text-books on the history of philosophy. And I am amazed now that I did not know it before, that there are so few who have even heard about it. Who knows, for instance, that an ordinary pack of playing-cards contains a profound and harmonious philosophical system? This is so entirely forgotten that it seems almost new.

I decide to write, to tell of all I have found. And at the same time I see that it is perfectly possible to make the ideas of this hidden thought agree with the data of exact knowledge, and I realise that the "fourth dimension" is the bridge that can be thrown across between the old and the new knowledge. And I see and find ideas of the fourth dimension in ancient symbolism, in the Tarot cards, in the images of Indian gods, in the branches of a tree, and in the lines of the human body.

I collect material, select quotations, prepare summaries, with the idea of showing the peculiar inner connection which I now see between methods of thinking that ordinarily appear separate and independent. But in the midst of this work, when everything is made ready, everything takes shape, I suddenly begin to feel a chill of doubt and weariness creeping over me. Well, one more book will be written, but even now, when I am only beginning to write it, I know how it will end. I know the limit beyond which it is impossible to go. The work stands still. I cannot make myself write about the limitless possibilities of knowledge when for myself I already see the limit. The old methods are no good; some other methods are necessary. People who think that something can be attained by their own efforts are as blind as those who are utterly ignorant of the possibilities of the new knowledge.

Work on the book is abandoned.

Months go by, and I become completely absorbed in strange experiments which carry me far beyond the limits of the known and possible.

Frightening and fascinating sensations. Everything becomes alive! There is nothing dead or inanimate. I feel the beating of the pulse of life. I " see " Infinity. Then everything vanishes. But each time I say to myself afterwards that this *has been* and, therefore, things exist that are different from the ordinary. But so little remains ; I remember so vaguely what I have experienced ; I can tell myself only an infinitesimal part of what has been. And I can control nothing, direct nothing. Sometimes *this* comes, sometimes it does not. Sometimes only horror comes, sometimes a blinding light. Sometimes a little remains in the memory, sometimes nothing at all. Sometimes much is understood, new horizons are disclosed, but only for a moment. And these moments are so short that I can never be certain whether I have seen anything or not. Light flares up and dies before I have time to tell myself what I have seen. And each day, each time, it becomes more and more difficult to kindle this light. It often seems that the first experiment gave me everything, that afterwards there has been nothing but a repetition of the same things in my consciousness, only a reflection. I know that this is not true and that each time I receive something new. But it is difficult to get rid of this thought. And it increases the sensation of helplessness that I feel in the face of the wall behind which I can look for a moment, but never long enough to account to myself for what I see. Further experiments only emphasise my powerlessness to get hold of the mystery. Thought does not grasp, does not convey, what is at times clearly felt. Thought is too slow, too short. There are no words and no forms to convey what one sees and knows in such moments. And it is impossible to fix these moments, to arrest them, to make them longer, more obedient to the will. There is no possibility of remembering what has been found and understood, and later repeating it to oneself. It disappears as dreams disappear. Perhaps it is all nothing but a dream.

Yet at the same time this is not so. I know it is not a dream. In these experiments and experiences there is a taste of reality which cannot be imitated and about which one cannot make a mistake. I know that *all this is there*. I have become convinced of it. *Unity exists*. And I know already that it is infinite, orderly, animated and conscious. But how to link " what is above " with " what is below " ?

I feel that a method is necessary. There is something which one must know before starting on experiments. And more and more often I begin to think that this method can be given only by those Eastern schools of Yogis and Sufis about which one reads and hears, *if such schools exist* and if they can be penetrated. My thought concentrates on this. The question of *school* and of a method acquires for me a predominant significance, though it is still not clear and is connected with too many fantasies and ideas based on very doubtful theories.   But one thing I see clearly, that alone, by myself, I can do nothing.

And I decide to start on a long journey with the idea of searching for those schools or for the people who may show me the way to them.

1912.

*   *   *

My way lay to the East.   My previous journeys had convinced me that there still remained much in the East that had long ceased to exist in Europe.   At the same time I was not at all sure that I should find precisely what I wanted to find.   And above all I could not say with certainty *what* exactly I should search for.   The question of " schools " (I am speaking, of course, of " esoteric " or " occult " schools) still contained much that was not clear.   I did not doubt that schools existed.   But I could not say whether it was necessary to assume the *physical* existence of such schools on earth.   Sometimes it seemed to me that true schools could only exist on another plane and that we could approach them only when in special states of consciousness, without actual change of place or conditions.   In that case my journey became purposeless.   Yet it seemed to me that there might be traditional methods of approach to esotericism still preserved in the East.

The question of schools coincided with the question of esoteric succession.   Sometimes it seemed to me possible to admit an uninterrupted historical succession.   At other times it seemed to me that only " mystical " succession was possible, that is, that the line of succession on earth breaks, goes out of our field of vision.   There remain only traces of it : works of art, literary memorials, myths, religions.   Then, perhaps only after a long interval of time, the same causes which once created esoteric thought begin to work once more, and once more there begins the process of " collecting knowledge ", *schools* are created and the ancient teaching emerges from its hidden form.   This would mean that during the intermediary period there could be no full or rightly organised schools, but only imitation schools or schools that preserve the letter of the old law, petrified in fixed forms.

However, this did not deter me.   I was ready to accept whatever the facts which I hoped to find should show me.

There was yet another question which occupied me before my journey and during the first part of it.

Should one and can one try to do something here and now with an obviously insufficient knowledge of methods, ways and possible results ?

In asking this I had in mind various methods of breathing, dieting, fasting, exercises of the attention and imagination and, above all, of overcoming oneself at moments of passivity or lassitude.

In answering this question voices in me were divided :

"It does not matter what one does, only one has to do something," said one voice ; "but one should not sit and wait for something to come to one of itself."

"The whole point is precisely to do nothing," said another voice, "until one knows surely and definitely what should be done to attain a definite aim. If one begins to do something without knowing exactly what is necessary for what object, this knowledge will never come. The result will be the 'work on oneself' of various 'occult' and 'theosophical' books, that is, make-believe."

And listening to these two voices within me I was unable to decide which of them was right.

Ought I to try or ought I to wait ? I understood that in many cases it was useless *to try*. How can one *try* to paint a picture ? How can one *try* to read Chinese ? One must first study and know, that is, be able to do it. At the same time I realised that in these last arguments there was much desire to evade difficulties or at least to postpone them. However, the fear of amateurish attempts at " work on oneself " outweighed the rest. I said to myself that in the direction I wanted to go it was impossible to go blindly, that one must see or know where one was going. Besides, I did not even wish for any changes in myself. I was going in search of something. If in the midst of this process of search I myself began to change, I should perhaps be satisfied with something quite different from what I wanted to search for. It seemed to me then that this is precisely what often happens to people on the road of " occult " search. They begin to try various methods on themselves and put so much expectation, so much labour and effort, into these attempts that in the end they take the subjective results of their efforts for the results of their search. I wanted to avoid this at all costs.

But a quite different and almost unexpected aim to my journey began to outline itself from the very first months of my travels.

In almost every place I came to, and even during the journey, I met people who were interested in the same ideas that interested me, who spoke the same language as I spoke, people between whom and myself there was instantly set up an entirely distinctive understanding. How far this special understanding would lead, of course I was unable to say at that time, but in the conditions and with the material of ideas I then possessed, even such understanding seemed almost miraculous. Some of these people knew one another, others did not. And I felt that I was establishing a link between them, was, as it were, stretching out a thread which, according to the original plan of my journey, should go round the world. There was something which drew me and which was full of significance in these encounters. To every new man I met I spoke of others I had met earlier, and sometimes I knew beforehand people I was to meet later.

St. Petersburg, London, Paris, Genoa, Cairo, Colombo, Galle, Madras, Benares, Calcutta, were connected by invisible threads of common hopes and common·expectation. And the more people I met, the more this side of my journey took hold of me. It was as though there grew out of it some secret society, having no name, no form, no conventional laws, but closely connected by community of ideas and language. I often thought of what I myself had written in *Tertium Organum* about people of a "new race". And it seemed to me that I had not been far from the truth, and that there is actually carried on the process of the formation, if not of a new race, at least of some new category of men, for whom there exist different values than for other people.

In connection with these thoughts I again came to the necessity of putting in order and arranging systematically that which among the whole of our knowledge leads to "new facts". And I decided that after my return I would resume the abandoned work on my book, but with new aims and with new intentions.

At the same time I began to make certain connections in India and in Ceylon, and it seemed to me that in a short time I should be able to say that I had found concrete facts.

But there came one brilliant sunny morning when, on my way back from India, I stood on the deck of the steamer going from Madras to Colombo and rounding Ceylon from the south. This was the third time I had approached Ceylon, during this period, on every occasion from a different direction. The flat shore with blue hills in the distance revealed simultaneously what could never be seen when one was there on the spot. Through my glasses I could see the toy railway going south and all at once several toy stations, which appeared to be almost side by side. I even knew their names : Kollupitiya, Bambalapitiya, Wellawatta, and others.

The approach to Colombo stirred me. I was to know there :

first, whether I should again find the man I had met before my last trip to India and whether he would repeat the proposal he had made me regarding my meeting certain Yogis, and secondly where I should go next : should it be back to Russia, or further on to Burma, Siam, Japan and America.

But I was not expecting what actually met me.

The first word I heard on landing was : *war*.

There began then strange muddled days. Everything was thrown into confusion. But I already felt that my search in one sense was ended and I understood then why I had all the time felt that it was necessary to hurry. A new cycle was beginning. And it was as yet impossible to say what it would be like and to what it would lead. One thing only was clear from the first, that what was possible yesterday became impossible to-day. All the mud was rising from the bottom of life. All the cards became mixed. All the threads were broken.

There remained only what I had established for myself. Nobody could take that from me. And it alone could lead me further.

1914-1930.

CHAPTER I
A NEW MODEL OF THE UNIVERSE

I

Question of the form of the universe—History of the question—Geometrical and physical space—Doubt as to their identity—The fourth coordinate of physical space —Relation of physical sciences to mathematics—Old and new physics—The basic principles of old physics—Space taken as separate from time—The principle of the unity of laws—The principle of Aristotle—Undefined quantities in old physics—The method of division used instead of definition—Organised and non-organised matter—Elements —Molecular motion—Brownian movement—The principle of the conservation of matter—Relativity of motion—Measurement of quantities—Absolute units of measure —Law of universal gravitation—Actio in distans—Æther—Hypotheses of light— Michelson-Morley experiment—Velocity of light as a limiting velocity—Lorentz's conclusions—The quantum theory—Ponderability of light—Mathematical physics—The theory of Einstein—Contraction of moving bodies—The special and the general principles of relativity—Four-dimensional continuum—Amended and supplemented geometry according to Einstein—Relation of theory of relativity to experience—The " mollusc " of Prof. Einstein—Finite space—Two-dimensional spherical space—Prof. Eddington on space—On the study of the structure of radiant energy—Old Physics and New Physics.

AT every attempt to study the world and nature man inevitably finds himself confronted with a series of definite questions to which he is unable to give direct answers. But upon his recognition or non-recognition of these questions, upon his way of formulating them, and upon his attitude towards them, depends the whole further process of his thinking about the world and, consequently, about himself.

The most important of these questions are the following :

1. *What form has the world ?*
2. *Is the world a chaos or a system ?*
3. *Did the world come into being accidentally, or was it created according to plan ?*

And strange though it may appear at the first glance, one or another solution of the first question, that of the form of the world, actually determines the possible solution both of the second and of the third questions.

If the questions as to whether the world is a chaos or a system, and whether the world came into being accidentally or was created according to plan, are solved without being preceded by a definition of the form of the world, and do not result from such a definition, those solutions lack weight, demand " faith " and fail to satisfy the

21

mind. It is only when the answers to these questions are derived from the definition of the form of the world that they can be suffi ciently exact and complete.

It is not difficult to prove that the predominating general philo- sophies of life of our time are based on such solutions of these three fundamental questions as might have been considered scientific during the 19th century The discoveries of the 20th century and even those of the end of the 19th century have not as yet affected ordinary thought or have affected it very little.

And it is not difficult to prove that all further questions concerning the world, the development and elaboration of which constitute the object of scientific, philosophical and religious thought, arise from these three fundamental questions.

But in spite of its predominant importance, the question of the form of the world has comparatively seldom arisen independently, being usually included in other problems, cosmogonical, cosmological, astronomical, geometrical, physical and other. The average man would be greatly surprised if he were told that the world may have a form. For him the *world* has no form.

Yet in order to understand the world one must be able to build some model of the universe, however imperfect. Such a model of the world, such a model of the universe, cannot be built without a definite conception of the form of the universe. To make a model of a house one must know the form of the house, to make a model of an apple, one must know the form of the apple.

Therefore, before passing to principles upon which a new model of the universe can be built, we must examine, though only summarily, the history of the question as to the form of the world, the present state of this question in science, and the " models " which have been built up to the present day.

The ancient and mediæval cosmogonical and cosmological concep- tions of exoteric systems (which alone became known to science) were never very clear or very interesting. Moreover, the universe they pictured was a very small universe, much smaller than the modern astronomical world. I shall not therefore speak of them.

Our study of different views of the question concerning the form of the world will begin from the moment when astronomical and physico-mechanical systems freed themselves from the idea that the earth is the centre of the world. The period in question embraces several centuries. But actually we shall occupy ourselves only with the last century, almost precisely from the end of the first quarter of the 19th century.

By that time the sciences which studied the world of nature had long been divided and stood then in the same relation to one another in which they stand now, or stood at any rate quite recently.

Physics studied phenomena in matter around us.

Astronomy studied the "movements of celestial bodies".

Chemistry endeavoured to penetrate the mystery of the structure and composition of matter.

These three physical sciences based their conceptions of the form of the world entirely upon the geometry of Euclid. Geometrical space was taken as physical space. No difference was distinguished between them, and space was taken apart from matter, just as a box and its capacity may be examined independently of its contents.

Space was understood as an "infinite sphere". The infinite sphere was geometrically determined only from the centre, that is, from any point, by three radii at right angles to one another. And an infinite sphere was regarded as entirely similar in all its physical properties to a finite sphere.

The question of the non-correspondence of geometrical, that is, of Euclidean, three-dimensional space (whether infinite or finite) on the one hand with physical space on the other hand arose only very occasionally and did not interfere with the development of physics in the directions which were possible to it.

It was only about the end of the 18th and the beginning of the 19th century that the idea of this non-correspondence and the doubt as to the correctness of identifying physical space with geometrical space became so insistent that it was no longer possible to pass them over in silence.

This doubt was aroused, first: by attempts at a revaluation of geometrical values, that is, attempts either to *prove* the axioms of Euclid, or to prove their incorrectness; and second: by the very development of physics, or more exactly of mechanics, that is, the part of physics dealing with motion, for this development led to the conviction that physical space could not be housed in geometrical space and continually reached beyond it. Geometrical space could be taken as physical space only by closing the eyes to the fact that in geometrical space everything is immovable, that it contains no *time* necessary for motion, and that the calculation of any figure resulting from motion, such as a screw, for instance, requires four coordinates.

Later on, the study of phenomena of light, electricity and magnetism, and also the study of the structure of the atom, necessitated a similar broadening of the concept of space.

The result of purely geometrical speculations concerning the correctness or incorrectness of the axioms of Euclid was twofold.

On the one hand a conviction arose that geometry was a purely speculative science, dealing solely with principles and entirely completed, which could neither be added to nor altered; also a science which could not be applied to all the facts that are met with, which is true only under certain definite conditions, but within those conditions is perfectly reliable and irreplaceable by anything else. On the other hand there arose a certain disappointment in the geometry of Euclid and a desire to remodel it, to rebuild it on a new basis, to broaden it, to make it a physical science which could be applied to all the facts that are met with, without the necessity for arranging these facts in an artificial order. The first view on geometry was right; the second was wrong, but this second attitude can be said ment. But I shall revert to this later.

Kant's ideas of categories of space and time taken as categories of perception and thought have never entered into scientific, i.e. physical thought, in spite of certain later attempts to introduce them into physics. Scientific (physical) thought proceeded apart from philosophical and psychological thought. And scientific thought always took time and space as having an objective existence outside us. And in virtue of this it was always considered possible to express their relations mathematically.

But the development of mechanics and other branches of physics led to the necessity for recognising a fourth coordinate of space in addition to the three fundamental coordinates : length, breadth and height. And the idea of the fourth coordinate or the fourth dimension of space gradually became more and more inevitable, though for a long time it remained a kind of " taboo ".

Material for the construction of new hypotheses of space remained in the works of the mathematicians : Gauss, Lobatchevsky, Saccheri, Bolyai and especially Riemann, who in the fifties of the 19th century was already considering the question of the possibility of a totally new understanding of space. There were no serious attempts at a psychological study of the problem of space and time. The idea of the fourth dimension remained for a long time shelved, and by specialists was regarded as purely mathematical and by non-specialists as mystical or occult.

But if we start from the moment of the appearance of this idea at the beginning of the 19th century and make a brief survey of the development of scientific thought from that moment up to the present day, it may help us to understand the course which the further development of the idea may take. At the same time we may see what this idea tells us or can tell us in regard to the fundamental problem of the form of the world.

The first and essential question which arises at this point is that of the relation of the physical sciences to mathematics. From the ordinary point of view it is taken as an admitted fact that mathematics studies the relation of quantities in the same world of things and phenomena as that studied by the physical sciences. From this follow two more propositions : first, that every mathematical proposition must have a physical equivalent, though it may still be undiscovered at the given moment; and second, that every physical phenomenon can be expressed mathematically.

As a matter of fact neither of these propositions has any foundation whatever, and the acceptance of them as axioms arrests the progress of thought along the very lines where progress is most necessary. But this will be dealt with later.

In the discussions which follow of all the physical sciences we shall examine only physics proper. And in physics we shall have to pay most attention at first to mechanics : for since about the middle of the 18th century mechanics has assumed a predominant position in physics ; so much so that until quite recently it was considered both possible and probable that a means would be found of interpreting all physical phenomena as mechanical phenomena, that is, as phenomena of motion. Some scientists even went much further in this direction and, not content with admitting the possibility of finding a means of interpreting physical phenomena as phenomena of motion, asserted that this means had already been found and that it explained not only physical phenomena, but also phenomena of life and thought.

At present one often meets with a division of physics into *old* and *new*, and in its chief lines this division may be accepted. But it should not be understood too literally.

I will now try to make a brief survey of the fundamental ideas of old physics which led to the necessity for building " new physics ", which has unexpectedly destroyed old physics ; and then I will come to the ideas of new physics which lead to the possibility of building a " new model of the universe ", which destroys new physics just as new physics destroyed old physics.

Old physics lasted until the discovery of the electron. But even the electron was conceived by old physics as existing in the same artificial world, governed by Aristotelian and Newtonian laws, in which it studied visible phenomena ; in other words, the electron was accepted as existing in the same world in which our bodies and other objects commensurable with them exist. Physicists did not understand that the electron belongs to another world.

Old physics was based on certain immovable foundations. The

space and time of old physics possessed very definite properties. First of all, they could be examined and calculated *separately*, i.e. the being of a thing in space in no way affected or touched its being in time. Further, there was one space for all that exists, and all that occurred in this space. Time also was one for all that exists and was measured always and for everything by one scale. In other words, it was considered possible to measure with one measure all movements possible in the universe.

The corner-stone of the whole understanding of the laws of the universe was the principle of Aristotle concerning the unity of laws in the universe.

This principle in its modern meaning can be formulated in the following way : in the whole of the universe and under all possible conditions the laws of nature must be identical; in other words, a law which has been established at one place in the universe must hold good at any other place in the universe. On the basis of this, science, in studying phenomena on the earth and in the solar system, assumed the existence of the same phenomena on other planets and in other solar systems.

This principle, attributed to Aristotle, in reality was certainly never understood by him in the sense which it had acquired in our times. The universe of Aristotle differed greatly from the universe as we conceive it. The thinking of the people of Aristotle's time differed greatly from the thinking of the people of our time. Many fundamental principles and many starting-points of thought, which we can accept as firmly established, had to be proved and established by Aristotle.

Aristotle endeavoured to establish the unity of laws in the sense of a protest against superstitions, against naïve magic, against naïve miracles, and so on. In order to understand the " principle of Aristotle " it is necessary to realise that he had still to prove that, if in general dogs cannot speak in human language, then one particular dog, say, in the island of Crete, *also* cannot speak ; or that if in general trees cannot move of themselves, then one particular tree *also* cannot move, and so on.

All this has of course been forgotten long ago, and from the principle of Aristotle there follows now the idea of the permanency of all physical concepts, such as motion, velocity, force, energy, etc. This means that what has once been regarded as motion always remains motion ; what has once been regarded as velocity always remains velocity, becoming " infinite velocity ".

In its primary meaning the " principle of Aristotle " is comprehensible and necessary and is nothing else than the law of the general consecutiveness of phenomena which belongs to logic. But in its modern meaning the " principle of Aristotle " is entirely wrong.

Even for new physics the concept of infinite velocity, which is exclusively based on the " principle of Aristotle ", has become impossible, and the " principle of Aristotle " must be completely abandoned before the planning of a new model of the universe becomes possible. I shall return to this question later.

In speaking of physics it is first of all necessary to analyse the very definition of the subject. According to the definition of text-books of this science, physics studies " matter in space and phenomena in this matter ".

And here we are at once faced with the fact that physics operates with undefined and unknown quantities which, for the sake of convenience (or owing to the difficulty of definition), are taken as known quantities and even as quantities requiring no definition.

There are formally distinguished in physics first: quantities requiring definition ; and second : " primary " quantities, the idea of which is considered to be inherent in all people. Prof. Chwolson in his *Text-book*[1] enumerates as primary quantities :

Extensions—linear-extension, area-extension and volume-extension, that is, the length of a straight line, the area of a portion of surface and the volume of a portion of space limited by surfaces ; extension being the measure of size and distance.

Time.

Velocity of uniform rectilinear motion.

These are naturally only examples, and Prof. Chwolson does not insist on the completeness of the list. In reality, the list is very long ; it includes space, infinity, matter, motion, mass and so on. In a word, practically all the concepts with which physics operates refer to undefined and undefinable quantities.

Of course in a great many cases it is impossible to avoid operating with unknown quantities, but it has become the traditional " scientific " method not to recognise anything as unknown, and to regard the " quantities " which elude definition as " primary ", the idea of which is inherent in everyone. The natural result has been that the whole of the vast edifice erected with tremendous labour has become artificial and unreal.

---

[1] As an example of a text-book on physics from which quotations can be made the author has taken Prof. O. D. Chwolson's *Text-book on Physics* (in Russian), (5th edition, in five volumes, Berlin, 1923). This book is neither better nor worse than any other text-book on physics and it can very well be taken as an example of text-book opinions and views. It is even better than many other books because of Prof. Chwolson's impartiality towards new theories.

In the definition of physics given above we meet with two undefined concepts : *space* and *matter*.

I have already referred to space in the preceding pages. As regards matter, Prof. Chwolson writes (*Text-book of Physics*, Vol. I, Introduction) :

In objectifying the cause of a sensation, that is, transferring this cause into a definite place in space, we conceive this space as containing something which we call *matter* or substance (page 2).

Further Prof. Chwolson says :

The use of the term " matter " was reserved exclusively for matter which is able to affect our organ of touch more or less directly (page 7).

Further, matter is divided into organised matter (of which living bodies and plants are composed) and non-organised matter.

This method of division instead of definition is applied in physics whenever definition is difficult or impossible, that is, in relation to all fundamental concepts. Later we shall often meet with this fact.

The difference between organised matter and non-organised matter is determined only by external characteristics. The origin of organised matter is admitted to be unknown. The transition of non-organised matter into organised matter may be observed (feeding, breathing), and it is admitted that such a transition takes place only in the presence and through the action of already existing organised matter. The mystery of the first transition remains hidden (Chwolson).

On the other hand we see that organised matter easily passes into non-organised matter, losing certain undefinable properties which we call *life*.

Many attempts have been made to regard organised matter as a particular case of non-organised matter, and to explain all the phenomena that take place in organised matter (i.e. phenomena of life), as a combination of physical phenomena. But these attempts, as well as attempts at the artificial creation of organised matter from non-organised matter, led to nothing and could neither create nor prove anything. In spite of this they left a very strong impress on general philosophies of life of a scientific kind, from the standpoint of which the " artificial creation of life " is recognised as not only possible but already partly attained. Followers of these philosophies regard the very name of *organic chemistry*, i.e. chemistry studying organised matter, as having merely a historical meaning, and define it as the " chemistry of carbon compounds ", although at the same time they cannot help admitting the special position of the chemistry of carbon compounds and its difference from general inorganic chemistry.

Non-organised matter is in its turn divided into simple matter and composite matter (this becomes the province of chemistry). Composite matter consists of a so-called chemical compound of several simple matters. Every matter can be divided into very small parts, called "particles". A *particle* is the smallest quantity of the given matter which is still capable of exhibiting at least the chief properties of this matter. The further divisions of matter, molecule, atom, electron, are so small that, taken separately, they do not possess any material properties, though this last fact is never sufficiently taken into account.

According to the most recent scientific ideas, non-organised matter consists of 92 elements or simple matters, though not all of them have as yet been discovered. There exists a hypothesis that the atoms of various elements are nothing but a combination of a certain number of atoms of hydrogen, which, in this case, is taken as fundamental or primary matter. Several theories exist concerning the possibility or the impossibility of the transition of one element into another. And in some cases such a transition has been established, which again contradicts the "principle of Aristotle".

Organised matter, or "carbon compounds", actually consists of four elements—hydrogen, oxygen, carbon and nitrogen, with a negligible admixture of other elements.

Matter possesses many properties, such as mass, volume, density, etc., which are in most cases only definable one relatively to another.

The temperature of a body is recognised as depending on the motion of molecules. Molecules are considered to be in perpetual motion; as physics defines it, they are constantly colliding and scattering in all directions and returning again. The greater the motion, the greater the shocks when they collide, the higher the temperature (Brownian movement).

If this were possible in reality, it would mean approximately that, for instance, several hundreds of motor-cars, swiftly moving in different directions in a large square of a big city, crash into one another every minute and disperse in various directions, remaining intact.

It is very curious that a *quick-motion* cinematographic film produces such an illusion. Moving objects lose their individuality and appear to collide and fly off in different directions or pass through one another.[1]

---

[1] The author once saw a quick-motion cinematograph picture of the Place de la Concorde, with motor-cars rushing from all directions and in all directions. And the impression was exactly as if the cars violently collided with one another every moment and flew apart, remaining all the time in the square and never leaving it.

How it can be that material bodies possessing mass, weight and very complicated structure and moving at great velocity, collide and scatter without being broken up and destroyed, is not explained by physics.

One of the most important conquests of physics was the establishment of the principle of the conservation of matter. This principle consists in the recognition of the fact that matter is never and in no physical or chemical conditions created anew, nor does it disappear. Its total amount remains constant. With the principle of conservation of matter are connected the principles established later, the principle of conservation of energy and the principle of conservation of mass.

Mechanics is the name given to the science of the motion of physical bodies and of the causes upon which the character of this motion may depend in various particular cases (Chwolson).

But, just as in the case of all other physical concepts, *motion* is not defined by physics. Physics only establishes the properties of motion —duration, velocity and direction in space, without which properties a phenomenon cannot be called motion.

The division and sometimes the definition of these properties take the place of the definition of motion itself, and the established characteristics of the properties of motion are referred to motion itself. Thus motion is divided into rectilinear and curvilinear, continuous and non-continuous, accelerating and retarding, uniform and variable.

The establishment of the principle of the relativity of motion led to a whole series of conclusions. The question arose : if the motion of a material point can be determined only by its position in relation to other bodies or points, then how is the motion to be determined if the other bodies or points also move ? And this question became especially complicated when it was established, not merely philosophically in the sense of πάντα ῥεῖ, but fully scientifically, with calculations and diagrams, that nothing is motionless in the universe, that everything without exception moves in one way or another, and that one motion can be established only relatively to another. But at the same time there were established cases of apparent immobility in motion. Thus it was established that separate component parts of a uniformly moving system of bodies maintain the same position in relation to one another as though the system were stationary. Thus, things inside a swiftly moving railway carriage behave in exactly the same way as when the carriage is standing still. And in the case of two or more moving systems of bodies, for instance in the case of two trains running on different tracks in the same direction or in different directions, it was established that their relative velocity is equal to the difference between, or the sum of, their respective velocities, according to the direction of the movement. Thus two trains approaching one another will approach with a velocity equal to the sum of their respective velocities. For one train overtaking another, the second train will run in a direction opposite to its own with a velocity equal to the difference between the respective velocities of the two trains. What is usually called the velocity of a train is the velocity which is ascribed to the train observed during its passage between two objects which are stationary for it, for instance between two stations, and so on.

The study of motion in general and of vibratory and undulatory movements in particular exercised a tremendous influence on the development of physics. Wave movements began to be regarded as a universal principle, and many attempts were made to reduce all physical phenomena to vibratory movements.

One of the fundamental methods of physics was the measurement of quantities.

The measurement of quantities was based upon certain principles, the most important of which was the principle of homogeneity, namely, quantities conforming to the same definition and differing from one another merely quantitatively were called homogeneous quantities and it was considered possible to compare them and measure one in relation to another. As to quantities which differed in definition, it was considered impossible to measure them one in relation to another.

Unfortunately, as has already been shown, there were very few *definitions* of quantities in physics and therefore definitions were generally replaced by their denominations.

But as mistakes in the denomination could always occur, and qualitatively different quantities could be named similarly, while qualitatively identical quantities could be named differently, physical measurements were unreliable. And the more so because here again the principle of Aristotle was felt—that is, a quantity once recognised as a quantity of a certain order always remained a quantity of that order. Different forms of energy passed into one another, matter passed from one state into another, but space (or a part of space) always remained space, time always remained time, motion always remained motion, velocity always remained velocity, and so on.

On these grounds it was agreed to regard as *incommensurable* only those quantities which were qualitatively different. Quantities which differed merely quantitatively were regarded as *commensurable*.

Continuing the subject of the measurement of quantities, it is necessary to point out that the units of measure used in physics are quite arbitrary and have no connection with the quantities that are measured. All the units of measure have only one thing in common— they are always borrowed from *elsewhere*. There is not a single case in which a characteristic of the given quantity itself is taken as the measure.

The artificiality of measures in physics has certainly never been a secret, and from the realisation of this artificiality follow attempts to establish, for instance, the measure of length *as a part of the meridian*. Naturally these attempts alter nothing, and parts of the human body, an " ell " or a " foot ", taken as units of measure, or a " metre ", i.e. a part of the meridian, are equally arbitrary. In reality things bear their own measure in themselves. And to find the measure of things is to understand the world. Physicists have dimly guessed this, though they have never succeeded in even approaching these measures.

Prof. Planck in 1900 (this really belongs to new physics) constructed a system of " absolute units ", taking as its basis " universal constants ", namely : first, the velocity of light in a vacuum ; second, the constant unit of gravitation ; third, a constant quantity which plays an important part in thermo-dynamics (energy divided by temperature) ; and fourth, a constant quantity which is called " action " (energy multiplied by time) and is the smallest possible quantity of action or its atom.

Using these quantities Planck obtains a system of units which he considers to be absolute and entirely independent of any arbitrary choice of man, and which he regards as *natural*.

Planck affirms that these quantities will retain their natural meaning so long as the laws of universal gravitation and of the propagation of light in a vacuum, and the two fundamental principles of thermodynamics, remain unchanged ; they will always be the same by whatever intelligent beings and by whatever methods they are determined.

But the law of universal gravitation and the law of the propagation of light in a vacuum are the two weakest points in physics, because in reality they are not what they are taken for. And therefore Planck's whole system of measures is very unreliable. What is interesting in it is not the result but only the principle, i.e. the recognition of the necessity for finding natural measures of things. The actual determination of absolute units of measure lies beyond the new model of the universe.

The law of universal gravitation was stated by Newton in his book : *Philosophiæ naturalis principia mathematica*, which was published in London in 1687. This law from the very beginning received two formulations : one scientific, the other popular.

The scientific formulation is as follows :

There are observed phenomena between two bodies in space which *can be described* by presuming that two bodies attract one another with a force directly proportional to the product of their masses and inversely proportional to the square of the distance separating them.

And the popular formulation is :

Two bodies *attract* one another with a force directly proportional to the product of their masses and inversely proportional to the square of the distance separating them.

In this second formulation the fact is entirely forgotten that the force of attraction is merely a fictitious quantity accepted only for a convenient description of phenomena. *And the force of attraction* is regarded as really existing both between the sun and the earth and between the earth and a falling stone.[1]

Prof. Chwolson writes in his *Text-book of Physics* : [2]

> The tremendous development of celestial mechanics, entirely based on the law of universal gravitation taken as a fact, made scientists forget the purely descriptive character of this law and see in it the final formulation of an actually existent physical phenomenon.

What is important in Newton's law is that it gives a very simple mathematical formulation which can be applied throughout the universe, and on the basis of which it is possible to calculate all movements, in particular the movements of celestial bodies, with astonishing accuracy. Newton certainly never established it as a fact that bodies are actually attracted by one another, nor did he establish *why* they are attracted or *through the mediation of what*.

How can the sun influence the motion of the earth through the void of space? How in general is it possible to conceive action through empty space? The law of gravitation does not give an answer to this question, and Newton himself was perfectly aware of this fact. Both he and his contemporaries, Huygens and Leibnitz, definitely gave warning against attempts to see in Newton's law the solution of the problem of action through empty space, and regarded this law merely as a *formula for calculation*. Nevertheless the tremendous achievements of physics and astronomy attained through the application of Newton's law caused scientists to forget this warning, and the opinion was gradually established that Newton had discovered the force of attraction.

---

[1] The most recent electro-magnetic theory of gravitational fields dogmatises the *second* view.

[2] Vol. I, p. 182.

Prof. Chwolson writes in his *Text-book of Physics* (Vol. I, pp. 181, 182, 183) :

> The term " actio in distans ", that is, " action at a distance ", desig-
> nates one of the most harmful doctrines that ever prevailed in physics
> and retarded its progress ; this doctrine admitted the possibility of
> immediate action by one object on another object at a certain distance
> from it, at a distance so great as to make immediate contact between
> the two impossible.
>
> In the first half of the 19th century the idea of action at a distance
> reigned supreme in science. Faraday was the first to point out the
> impossibility of the admission that a body should *without mediation*
> excite forces and produce motion at a point where that body is not
> situated. Leaving aside the question of universal gravitation, he
> turned his special attention to magnetic and electric phenomena and
> pointed out the supremely important part played in these phenomena
> by the *intervening medium* which fills the space between the bodies that
> appear to act upon one another without mediation. . . .
>
> At the present time the conviction that action at a distance should
> not be admitted in any domain of physical phenomena has obtained
> universal recognition.

But the old physics was able to abandon action at a distance only
after it had accepted the hypothesis of the *universal medium* or æther.
The acceptance of that hypothesis was equally necessary for the
theories of light and electric phenomena as they were understood by
old physics.

In the 18th century phenomena of light were explained by the
hypothesis of emission put forward by Newton in 1704. This hypothesis assumed that luminous bodies emit in all directions minute
particles of a special light-substance which travel through space
with tremendous velocity and, entering the eye, produce in it the
impression of light. In this hypothesis Newton developed the ideas
of the ancients. In Plato the expression, " light filled my eyes ",
is often found.

Later, mainly in the 19th century, when the attention of investigators
was drawn to those results of the phenomena of light which could
not be explained on the hypothesis of emission, another hypothesis
obtained wide recognition, namely, the hypothesis of undulatory
vibrations in æther. This hypothesis was first advanced by the
Dutch physicist Huygens in 1690, but for a long time it was not
accepted by science. Later on, investigations of the phenomena of
diffraction definitely turned the scale in favour of the hypothesis of
light waves as against the hypothesis of emission ; and the subsequent
work of physicists, mainly on the polarisation of light, for a time
gained general recognition for this hypothesis.

In this hypothesis the phenomena of light are explained as analogous to the phenomena of sound. Just as sound results from the vibration of particles of the sonant body and is propagated through the vibration of particles of the air or some other elastic medium, so, on this hypothesis, light results from the vibration of molecules of the luminous body and is propagated by means of vibrations in an exceedingly elastic æther which fills both interstellar space and the space between molecules.

During the 19th century the theory of vibrations gradually became the basis of the whole of physics. Electricity, magnetism, heat, light, even *life* and *thought* (purely dialectically, it is true), were explained by the theory of vibrations. And it cannot be denied that in the case of the phenomena of light and electro-magnetics the theory of vibrations gave remarkably convenient and simple formulæ for calculation. A whole series of remarkable discoveries and inventions was made on the basis of the theory of vibrations.

But the theory of vibrations required æther. Æther as a hypothesis was created for the explanation of very heterogeneous phenomena, and it was therefore endowed with strange and contradictory properties. It is omnipresent, it fills the whole universe, pervades all its points, all atoms and all interatomic space. It is continuous, it possesses perfect elasticity. Yet æther is so rarefied, thin and permeable that all earthly and heavenly bodies pass through it without meeting with perceptible resistance to their movement. Its rarity is so great that if æther were to be condensed into a liquid, the whole of its mass within the limits of the system of the Milky Way could be contained in one cubic centimetre.

At the same time Sir Oliver Lodge considers the density of æther to be approximately a *billion times* greater than the density of water. From the latter point of view the world proves to be composed of a solid substance—" æther "—which is millions of times denser than a diamond ; and matter, even the densest matter we know, is merely *empty space*, a bubble in the mass of æther.

Many attempts have been made to prove the existence of æther or to discover facts confirming its existence.

Thus it was recognised that the existence of aether would be established if it were once proved that a ray of light moving faster than another ray of light changes its character in a certain way.

It is a known fact that the pitch of a sound rises or falls as the hearer approaches or retreats from it (Doppler's principle). Theoretically this principle was considered applicable to light. This would have meant that a swiftly approaching or retreating object should change its colour (as the sound of an engine-whistle changes its pitch as it approaches or retreats). But owing to the structure of the eye and the speed of its perception it was impossible to expect that the eye would notice the change of colour even if such a change actually took place.

In order to establish the fact of the change of colour it was necessary to have recourse to the spectroscope, that is, to decompose a ray of light and observe every colour of the spectrum separately.

These experiments gave no positive results whatever and to prove the existence of the æther by them was not possible.

In order to settle once and for all the question of the existence or the non-existence of the æther the American scientists Michelson and Morley, in the middle eighties of the last century, began a whole series of experiments assisted by special apparatus invented by themselves.[1]

The apparatus was mounted on a stone slab fixed upon a wooden float revolving in a tank filled with mercury, and made one full revolution in six minutes. A ray of light from a special lamp fell on mirrors attached to the revolving float and partly passed through them and partly was reflected, one half going in the direction of the movement of the earth and the other at right angles to it. This means that in accordance with the plan of the experiment one half of the ray moved with the normal speed of light and the other with the speed of light *plus* the speed of the rotation of the earth. At the union of the divided ray, there should have appeared, according to the plan of the experiment, certain light phenomena resulting from a difference in speed and showing the relative movement between the earth and the æther, that is, indirectly proving the existence of the æther.

Observations were made over a long period at all times of the day and night, and nothing was discovered.

From the standpoint of the original problem it was necessary to recognise that the experiment failed. But it disclosed another phenomenon, possibly much more significant than that which it attempted to establish. This was the fact that the speed of a ray of light cannot be increased. The ray of light moving with the earth differed in no way from the ray of light moving at right angles to the direction of the movement of the earth in its orbit.

It was necessary to recognise as a *law* that the velocity of a ray of light is a constant and limiting quantity, which cannot be increased. And this, in its turn, explained why Doppler's principle was inapplicable to phenomena of light. At the same time it established the fact that the general law of the composition of velocities, which

---

[1] For the detailed description of the experiment of Michelson and Morley, see the *American Journal of Science* (Third Series), 1887, Vol. 34, pp. 333 *et seq.*

was the basis of mechanics, could not be applied to the velocity of light.

In his book on relativity, Prof. Einstein explains that if we imagine a train moving at the rate of 30 kilometres a second, i.e. with the velocity of the movement of the earth, and a ray of light overtaking or meeting it, then the composition of velocities will in this case be impossible. The velocity of light will not be increased by the addition to it of the velocity of the train and will not be decreased by the subtraction from it of the velocity of the train.

At the same time it was established that no existing instruments or means of observation can *intercept a moving ray*. In other words, it is never possible to catch the end of a ray which has not yet reached its destination. In theory we may speak of rays which have not yet reached a certain point, but in practice we are unable to observe such rays. Consequently, for us, with our means of observation, the propagation of light is virtually instantaneous.

At the same time the physicists who analysed the results of the Michelson-Morley experiment explained its failure by the presence of new and unknown phenomena resulting from great velocities.

The first attempts to solve this question were made by Lorentz and Fitzgerald. *The experiment could not succeed*, was Lorentz's formulation of his propositions, for every body moving in æther *itself* undergoes deformation, namely, for an observer at rest it contracts in the direction of the motion. Basing his reasonings on the fundamental laws of mechanics and physics, he showed by means of a series of mathematical constructions that the Michelson-Morley installation necessarily suffered a contraction and that the amount of the contraction was exactly such as to counterbalance the displacement of the light waves consequent upon their direction in space, and thus to annul the results of the difference in velocity of the two rays.

Lorentz's conclusions as to the presumed contraction of a moving body gave rise in their turn to many explanations, and one of these explanations was put forward from the point of view of Prof. Einstein's special principle of relativity.

But this relates to the new physics.

The old physics was indissolubly connected with the theory of vibrations.

The new theory, which came to replace the mechanical theory of vibrations, was the theory of the atomic structure of light and electricity, taken as independently existing matters composed of *quanta*.

The new teaching, says Prof. Chwolson,[1] appears to be a return to the Newtonian emission theory although considerably altered.

[1] Vol. I, p. 9.

This new teaching is far from being completed. And its most important part, the *quantum* itself, still remains undefined. What a quantum is cannot be defined by new physics.

The theory of the atomic structure of light and electricity entirely altered the view on electrical and light phenomena. Science has ceased to see the fundamental cause of electrical phenomena in special states of æther and has returned to the old doctrine which admitted electricity to be a kind of substance which has real existence.

The same thing has happened with light. According to modern theories, light is a stream of minute particles rushing through space at the rate of 300,000 kilometres a second. They are not the corpuscles of Newton, but a special kind of *matter-energy*, formed by electromagnetic vortices.

The materiality of the light stream was established by the experiments of Prof. Lebedeff of Moscow. Prof. Lebedeff proved that light has weight, that is to say, that light when falling on bodies produces a mechanical pressure on them. It is characteristic that at the beginning of his experiments to determine the weight of light Prof. Lebedeff based them on the theory of the vibrations of the æther. This shows how the old physics confuted itself.

Prof. Lebedeff's discovery was very important for astronomy; for instance it explained certain phenomena which had been observed at the passing of the tail of a comet near the sun. But it was chiefly important for physics, as it supplied a further confirmation of the unity of the structure of radiant energy.

The impossibility of proving the existence of the æther, the establishment of the limiting and constant velocity of light, new theories of light and electricity and, above all, the study of the structure of the atom, indicated the most interesting lines of the development of the new physics.

Another part of new physics has developed from that particular formation of physics which was called mathematical physics. According to the definition which was given to it, mathematical physics usually started from some fact confirmed by experiment and expressing a certain orderly connection between phenomena. It enveloped this connection in a mathematical form and further, as it were, transformed itself almost into pure mathematics and began to elaborate, exclusively by means of mathematical analysis, those consequences which followed from the basic proposition (Chwolson).

Thus it is presumed that the success or unsuccess of the conclusions of mathematical physics might depend upon three factors : first, on the correctness of the definition of the fundamental fact, second, on the correctness of its mathematical expression, and third, on the correctness of the subsequent mathematical analysis.

There was a time when the importance of mathematical physics was greatly exaggerated, writes Prof. Chwolson (Vol. I, p. 13).

It was expected that it was precisely mathematical physics which should have served the principal course of the development of physics as a science. This, however, is quite erroneous. In the deductions of mathematical physics there are a great number of essential defects. In the first place, in almost every case it is only in the first rough approximation that they correspond with the results of direct observation. This is caused by the fact that the premises of mathematical physics can be considered sufficiently exact only within the narrowest limits : moreover these premises generally disregard a whole series of collateral circumstances the influence of which outside these narrow limits cannot be neglected. Therefore, the deductions of mathematical physics correspond to ideal cases, which cannot be practically realised and are often far removed from actuality.

And further :

It should be added that the methods of mathematical physics make it possible to solve special problems in hardly any but the simplest cases, especially so far as the *form* of the body is concerned. But practical physics cannot limit itself to these cases and is continually faced with problems which mathematical physics is incapable of solving. Moreover the results of the deductions of mathematical physics are often so complicated that their practical application proves to be impossible.

In addition to this should be mentioned yet another very characteristic property of mathematical physics, namely, that as a rule its deductions cannot be formulated otherwise than mathematically, and lose all their meaning and importance if an attempt be made to interpret them in the language of facts.

The new physics which developed from mathematical physics possesses many of the properties of the latter.

Prof. Einstein's theory of relativity is a separate chapter in new physics, which has developed from mathematical physics. It is wrong to identify the theory of relativity with new physics as is done by some followers of Prof. Einstein. New physics can exist without the theory of relativity. But for us, from the standpoint of the construction of a model of the universe, the theory of relativity is of great interest because it deals before anything else with the fundamental question of the form of the world.

There exists an enormous literature devoted to the exposition, explanation, popularisation, criticism and elaboration of the principles of Einstein, but, owing to the close relationship between the theory of relativity and mathematical physics, deductions from this theory are difficult to formulate logically. And the fact must be accepted that neither Prof. Einstein himself nor any of his numerous followers and interpreters have succeeded in explaining the meaning and essence of his theories in a clear and comprehensible way.

One of the first reasons for this fact is pointed out by Mr. Bertrand Russell in his popular book, *The A B C of Relativity*. He writes that the name " the theory of relativity " misleads people, and that a tendency to prove that *everything is relative* is generally ascribed to Prof. Einstein, while in reality he endeavours to discover and establish *that which is not relative*. And it would be still more correct to say that Prof. Einstein endeavours to establish the relation between what is relative and what is not relative.

Further Prof. Chwolson, in his *Text-book of Physics*, writes of the theory of relativity (Vol. V, p. 350):

The foremost place in Einstein's theory of relativity is occupied by a perfectly new and, at first glance, incomprehensibly strange conception of time. Much effort and prolonged work on oneself are needed to become used to it. But it is infinitely more difficult to accept the numerous consequences which follow from the principle of relativity and affect all branches of physics without exception. Many of these consequences obviously contradict what is usually, though often without adequate motive, called " common sense ". Some of these may be called the paradoxes of the new doctrine.

Einstein's ideas about time may be formulated as follows:

Each of two systems moving relatively to each other has in fact its own time, perceived and measured by an observer moving with the particular system.

The concept of simultaneity in the general sense does not exist. Two events which occur at different places may appear simultaneous to an observer at one point, whereas for an observer at another point they may occur at different times. It is possible that for the first observer the same phenomenon may occur earlier, and for the second, later (Chwolson).

Further, of the ideas of Prof. Einstein, Prof. Chwolson singles out the following:

The æther does not exist.

The concept of space, taken separately, has no meaning whatever. Only co-existence of space and time makes reality.

Energy possesses inert mass. Energy is an analogue of matter, and the transformation of what we call the mass of ponderable matter into the mass of energy, and vice versa, is possible.

It is necessary to distinguish the geometrical form of a body from its kinetic form.

The last points to a definite connection between Einstein's theory and the supposition of Fitzgerald and Lorentz as to the lengthwise contraction of moving bodies. Einstein accepts this supposition, although he says that he bases it on other principles than those of Fitzgerald and Lorentz, namely, on the special principle of relativity. At the same time the theory of the lengthwise contraction of bodies, deduced not from facts but from Lorentz's transformations, becomes the necessary foundation of the theory of relativity.

In making use exclusively of Lorentz's transformations, Einstein affirms that a rigid rod moving in the direction of its length is shorter than the same rod when it is in a state of rest, and the more quickly such a rod moves, the shorter it becomes. A rod moving with the velocity of light would lose its third dimension. It would become a cross-section of itself.

Lorentz himself affirmed that an electron actually disappeared when moving with the velocity of light.

These affirmations cannot be proved, since the contractions, even if they really occur, are too negligible with all possible velocities. A body moving with the velocity of the earth, i.e. 30 kilometres a second, must, according to the calculations of Lorentz and Einstein, undergo contraction by $\dfrac{1}{200,000}$ of its length; that is, a body 200 metres long would contract by 1 millimetre.

Further it is interesting to note that the supposition as to the *contraction of a moving body* radically contradicts the principle established by new physics, of the *increment of energy and mass in the moving body*. This latter principle is perfectly correct, although it has remained unelaborated.

As will be seen later, this principle, in its full meaning, which had not yet been revealed in new physics, is one of the foundations of the new model of the universe.

Passing to Einstein's own exposition of his fundamental theory, we see that it consists of two " principles of relativity ", the " special principle " and the " general principle ".

The " special principle of relativity " is supposed to establish the possibility of examining together and on the basis of a general law facts of the general relativity of motion which appear from the ordinary point of view to be contradictory, or to speak more accurately, the fact that all velocities are relative and that at the same time the velocity of light is non-relative, limiting and " maximal ". Einstein finds a way out of the difficulty created by all this, first : by understanding time itself, according to the formula of Minkovsky, as an imaginary quantity resulting from the relation of the given velocity to the velocity of light ; second : by making a whole series of altogether arbitrary assumptions on the border line of physics and geometry ; and third : by replacing direct investigations of physical phenomena and observations of their correlations by purely mathematical operations with Lorentz's transformations, the results of which show, in his opinion, the laws governing physical phenomena.

The " general principle of relativity " is introduced where it becomes necessary to make the idea of the infinity of space-time agree with the laws of the density of matter and the laws of gravitation in the space accessible to observation.

To put it briefly, the " special " and the " general " principles of relativity are necessary for agreement between contradictory theories on the border line of old and new physics.

The fundamental tendency of Einstein is to regard mathematics, geometry and physics as one whole.

The principle is certainly quite correct ; the three *ought to* constitute one. But " *ought to constitute* " does not mean that they *do constitute*.

The confusion of these two concepts is the chief defect of the theories of relativity.

In his book *The Theory of Relativity* Prof. Einstein writes :

> Space is a three-dimensional continuum. . . . Similarly the world of physical phenomena which was briefly called " world " by Minkovsky is naturally four-dimensional in the space-time sense. For it is composed of individual events, each of which is described by four numbers, namely, three space-coordinates and a time-coordinate. . . .
> That we have not been accustomed to regard the world in this sense as a four-dimensional continuum is due to the fact that in physics, before the advent of the theory of relativity, time played a different and more independent rôle, as compared with the space-coordinates. It is for this reason that we have been in the habit of treating time as an independent continuum. As a matter of fact, according to classical mechanics, time is absolute, i.e., it is independent of the position and the condition of motion of the system of coordinates. . . .
> The four-dimensional mode of consideration of the " world " is natural on the theory of relativity, since according to this theory time is robbed of its independence.

....

But the discovery of Minkovsky which was of importance for the formal development of the theory of relativity, does not lie here. It is to be found rather in the fact of his recognition that the four-dimensional space-time continuum of the theory of relativity, in its most essential formal properties, shows a pronounced relationship to the three-dimensional continuum of Euclidean geometrical space. In order to give due prominence to this relationship, however, we must replace the usual time coordinate t by an imaginary magnitude $\sqrt{-1}$. ct proportional to it. Under these conditions the natural laws satisfying the demands of the (special) theory of relativity assume mathematical forms, in which the time coordinate plays exactly the same rôle as the three space-coordinates. Formally these four coordinates correspond exactly to the space-coordinates in Euclidean geometry.[1]

The formula $\sqrt{-1}ct$ means that the time of every event is taken not simply by itself, but as an imaginary quantity in relation to the velocity of light, i.e. that a purely physical concept is introduced into the presumed " meta-geometrical " expression.

The time-duration t is multiplied by the velocity of light c and by the square root of minus one, $\sqrt{-1}$, which without changing the magnitude makes it an imaginary quantity.

This is quite clear. But what is necessary to note in relation to the passage quoted above is that Einstein regards Minkovsky's "world" as a development of the theory of relativity, whereas in reality the special principle of relativity *is built on the theory of Minkovsky*. If we suppose that the theory of Minkovsky is derived from the principle of relativity, then again, just as in the case of the theory of Fitzgerald and Lorentz relating to the lengthwise contraction of moving bodies, *it remains incomprehensible on what basis the principle of relativity is actually built*.

In any case, the building of the principle of relativity requires specially prepared material.

In the very beginning of his book Prof. Einstein writes that in order to make certain deductions from the observation of physical phenomena agree with one another it is necessary to revise certain *geometrical* concepts. "Geometry" means "land measuring", he writes.[2] " Both mathematics and geometry owe their origin to the need to know something of the properties of real things." On the basis of this, Prof. Einstein considers it possible to "supplement geometry", that is, for instance, to replace the concept of *straight lines* by the concept of *rigid rods*. Rigid rods are subject to changes under the influence of temperature, pressure, etc. ; they can expand and contract. All this must of course greatly alter "geometry".

---

[1] A. Einstein, *Relativity, the Special and the General Theory*. Translated by R. W. Lawson. 4th edition. Methuen & Co., London, pp. 55, 56, 57.
[2] *On the Physical Nature of Space*.

Geometry which has been supplemented in this way is obviously a natural science, says Einstein, and is to be treated as a branch of physics.[1]

I attach special importance to the view on geometry expounded here, because without it it would have been impossible to construct the theory of relativity.[2]

....

Euclidean geometry must be abandoned.[2]

The next important point in Einstein's theory is his justification of the mathematical method that he applies.

Experience has led to the conviction, he says, that, on the one hand, the principle of relativity (in the restricted sense) [3] holds true, and that on the other hand the velocity of the transmission of light in vacuo has to be considered to be a constant (*Relativity*, p. 42).

According to Prof. Einstein, the combination of these two propositions supplies the law of transformations for the four coordinates determining the place and the time of an event.

He writes :

Every general law of nature must be so constituted that it is transformed into a law of exactly the same form when, instead of the space-time variables of the original coordinate system, we introduce new space-time variables of another coordinate system. In this connection the mathematical relation between the magnitudes of the first order and the magnitudes of the second order is given by the Lorentz transformation. Or, in brief : General Laws of nature are co-variant with respect to Lorentz transformations (p. 42).

Einstein's assertion that the laws of nature are co-variant with Lorentz's transformations is the clearest illustration of his position. Starting from this point he considers it possible to ascribe to phenomena the changes which he finds in the transformations. This is precisely the method of mathematical physics which was condemned long ago, and which is mentioned by Prof. Chwolson in the passage quoted above.

In *The Theory of Relativity*, there is a chapter under the title " Experience and the Special Theory of Relativity."

To what extent is the special theory of relativity supported by experience ? This question is not easily answered, writes Prof. Einstein (p. 49).

....

The special theory of relativity has crystallised out from the Maxwell-

[1] *On the Physical Nature of Space.*
[2] Ibid.
[3] I.e. the principle of the relativity of velocities in classical mechanics.

Lorentz theory of electro-magnetic phenomena. Thus all facts of experience which support the electro-magnetic theory also support the theory of relativity (p. 49).

Prof. Einstein feels very acutely the necessity of facts for establishing his theories on firm ground. But he succeeds in finding these facts only in respect of invisible quantities—electrons and ions. He writes :

Classical mechanics required to be modified before it could come into line with the demands of the special theory of relativity. For the main part, however, this modification affects only the laws for rapid motions, in which the velocities of matter are not very small as compared with the velocity of light. We have experience of such rapid motions only in the case of electrons and ions ; for other motions the variations from the laws of classical mechanics are too small to make themselves evident in practice (p. 44).

Passing to the general theory of relativity, Prof. Einstein writes :

The classical principle of relativity, relating to three-dimensional space with the coordinate of time t (a real quantity) is violated by the fact of the constant velocity of light.

But the fact of the constant velocity of light is violated by the curvature of a ray of light in gravitational fields. This requires a new theory of relativity and a space, determined by Gaussian co-ordinates, applicable to non-Euclidean continua.

Gaussian coordinates differ from the Cartesian by the fact that they can be applied to any kind of space, independently of the properties of that space. They adapt themselves automatically to any space, whereas the Cartesian coordinates require a space of special definite properties, i.e. geometrical space.

In continuing the comparison of the special and the general theories of relativity Prof. Einstein writes :

The special theory of relativity has reference to domains in which no gravitational field exists. In this connection a rigid body in the state of motion serves as a body of reference, i.e. a rigid body the state of motion of which is so chosen that the proposition of the uniform rectilinear motion of " isolated " material points holds relatively to it (p. 98).

In order to make clear the principles of the general theory of relativity, Einstein takes the space-time domain as a disc uniformly rotating round its centre on its own plane. An observer situated on this disc regards the disc as being " at rest ". He regards the force acting upon him, and generally upon all bodies which are at rest in relation to the disc, as the action of the gravitational field.

The observer performs experiments on his circular disc with clocks
and measuring-rods. In doing so, it is his intention to arrive at exact
definitions for the signification of time and space data with reference
to the circular disc.

To start with, he places one of two identically constructed clocks
at the centre of the circular disc, and the other on the edge of the disc,
so that they are at rest relative to it (p. 80).

....

Thus on our circular disc, or, to make the case more general, in
every gravitational field, a clock will go more quickly or less quickly,
according to the position in which the clock is situated (at rest). For
this reason it is not possible to obtain a reasonable definition of time
with the aid of clocks which are arranged at rest with respect to the
body of reference. A similar difficulty presents itself when we attempt
to apply our earlier definition of simultaneity in such a case (p. 81).

....

The definition of the space coordinates also presents insurmount-
able difficulties. If the observer (moving with the disc) applies his
standard measuring-rod (a rod which is short as compared with the
radius of the disc) tangentially to the edge of the disc, then, . . . the
length of this rod will be less since moving bodies suffer a shortening
in the direction of the motion. On the other hand, the measuring-
rod will not experience a shortening in length, if it is applied to the
disc in the direction of the radius (p. 81).

....

For this reason non-rigid (elastic) reference-bodies are used, which
are as a whole not only moving in any way whatsoever, but which
also suffer alterations in form *ad lib.* during their motion. Clocks,
for which the law of motion is of any kind, however irregular, serve
for the definition of time. We have to imagine each of these clocks
fixed at a point on the non-rigid (elastic) reference-body. These clocks
satisfy only the one condition, that the " readings " which are observed
simultaneously on adjacent clocks (in space) differ from each other
by an infinitely small amount. This non-rigid (elastic) reference-body
which might appropriately be termed a " reference-mollusc ", is in the
main equivalent to a Gaussian four-dimensional coordinate system
chosen arbitrarily. That which gives the " mollusc " a certain com-
prehensibleness as compared with the Gauss coordinate system is
the (really unjustified) formal retention of the separate existence of the
space coordinates as opposed to the time coordinate. Every point
of the mollusc is treated as a space-point, and every material point
which is at rest relatively to it as at rest, so long as the mollusc is considered as reference-body. The general principle of relativity requires
that all these molluscs can be used as reference-bodies with equal right
and equal success in the formulation of the general laws of nature ;
the laws themselves must be quite independent of the choice of mollusc
(p. 99).

In respect of the fundamental question as to the form of the world
Einstein writes :

If we ponder over the question as to how the universe, considered as a whole, is to be regarded, the first answer that suggests itself is surely this : As regards space (and time) the universe is infinite. There are stars everywhere, so that the density of matter, although very variable in detail, is nevertheless on the average everywhere the same. In other words : However far we might travel through space, we should find everywhere an attenuated swarm of fixed stars of approximately the same kind and density (p. 105).

This view is not in harmony with the theory of Newton. The latter theory rather requires that the universe should have a kind of centre in which the density of the stars is a maximum, and that as we proceed outwards from this centre the group-density of the stars should diminish, until finally, at great distances, it is succeeded by an infinite region of emptiness. The stellar universe ought to be a finite island in the infinite ocean of space (pp. 105, 106).

The reason why an unbounded universe is impossible is that, according to the theory of Newton, the intensity of the gravitational field at the surface of a sphere filled with matter, even if this matter is of a very small density, would increase with increasing radius of the sphere, and would ultimately become infinite, which is impossible (p. 106).

The development of non-Euclidean geometry led to the recognition of the fact, that we can cast doubt on the infiniteness of our space without coming into conflict with the laws of thought or with experience (p. 108).

Admitting the possibility of similar conclusions Einstein describes the world of two-dimensional beings on a spherical surface.

In contrast to ours, the universe of these beings is two-dimensional ; but, like ours, it extends to infinity (p. 108).

This surface of the world of two-dimensional beings would constitute " space " for them. This space would possess very strange properties. If the spherical-surface beings were to draw circles in their " space ", that is, on the surface of their sphere, these circles would increase up to a certain limit, and would then begin to decrease.

The universe of these beings is finite and yet has no limits (p. 109).

Einstein comes to the conclusion that the spherical-surface beings would be able to determine that they are living on a sphere and might even find the radius of this sphere if they were able to examine a sufficiently great part of the surface.

But if this part is very small indeed, they will no longer be able to demonstrate that they are on a spherical " world " and not on a Euclidean plane, for a small part of a spherical surface differs only slightly from a piece of a plane of the same size (p. 110).

Thus if the spherical-surface beings are living on a planet of which the solar system occupies only a negligibly small part of the spherical universe, they have no means of determining whether they are living in a finite or an infinite universe, because the " piece of universe " to which they have access is in both cases practically plane, or Euclidean (p. 110).

....

To this two-dimensional sphere-universe there is a three-dimensional analogy, namely, the three-dimensional spherical space which was discovered by Riemann. Its points are likewise all equivalent. It possesses a finite volume which is determined by its " radius " (p. 111).

It is easily seen that the three-dimensional spherical space is quite analogous to the two-dimensional spherical surface. It is finite (that is of finite volume) and has no bounds (p. 112).

It may be mentioned that there is yet another kind of curved space, " elliptical space ". It can be regarded as a curved space in which the two " counterparts " are identical. . . . An elliptical universe can thus be considered to some extent as a curved universe possessing central symmetry (p. 112).

It follows from what has been said, that closed spaces without limits are conceivable. From amongst these, the spherical space (and the elliptical) excels in its simplicity, since all points in it are equivalent. As a result of this discussion, a most interesting question arises for astronomers and physicists, and that is whether the universe in which we live is infinite, or whether it is finite in the manner of the spherical universe. Our experience is far from being sufficient to enable us to answer this question. But the general theory of relativity permits of our answering it with a moderate degree of certainty, and in this connection the difficulty mentioned earlier (from the point of view of the Newtonian theory) finds its solution (p. 112).

The structure of space according to the general theory of relativity differs from that generally recognised.

According to the general theory of relativity, the geometrical properties of space are not independent, but they are determined by matter. Thus we can draw conclusions about the geometrical structure of the universe only if we base our considerations on the state of the matter as being something that is known. We know from experience that . . . the velocities of the stars are small as compared with the velocity of transmission of light. We can thus as a rough approximation arrive at a conclusion as to the nature of the universe as a whole, if we treat the matter as being at rest (p. 113).

....

We might imagine that as regards geometry, our universe behaves analogously to a surface which is irregularly curved in its individual parts, but which nowhere departs appreciably from a plane : something like the rippled surface of a lake. Such a universe might fittingly be called a quasi-Euclidean universe. As regards its space it would be infinite. But calculation shows that in a quasi-Euclidean universe the average density of matter would necessarily be nil. Thus such a universe could not be inhabited by matter everywhere : it would present to us an unsatisfactory picture (p. 114).

If we are to have in the universe an average density of matter which differs from zero, however small may be that difference, then the universe cannot be quasi-Euclidean. On the contrary the results of calculation indicate that if matter be distributed uniformly, the universe would necessarily be spherical (or elliptical). Since in reality the detailed distribution of matter is not uniform, the real universe will deviate in individual parts from the spherical, i.e. the universe will be quasi-spherical. But it will be necessarily finite. In fact, the theory supplies us with a simple connection between the space-expanse of the universe and the average density of matter in it (p. 114).

The last proposition is treated in a somewhat different manner by Prof. A. S. Eddington in his book : *Space, Time and Gravitation*.

After mass and energy there is one physical quantity which plays a very fundamental part in modern physics, known as *Action*.[1] *Action* here is a very technical term, and is not to be confused with Newton's " Action and Reaction ". In the relativity theory in particular this seems in many respects to be the most fundamental thing of all. The reason is not difficult to see. If we wish to speak of the continuous matter present *at* any particular point of space and time, we must use the term *density*. *Density multiplied by volume in space gives us mass*, or what appears to be the same thing, *energy*. But from our space-time point of view, a far more important thing is density multiplied by a four-dimensional volume of space and time ; this is *action*. The multiplication by three dimensions gives mass or energy ; and the fourth multiplication gives mass or energy multiplied by time. Action is thus mass multiplied by time, or energy multiplied by time, and is more fundamental than either.

Action is the curvature of the world. It is scarcely possible to visualise this statement, because our notion of curvature is derived from surfaces of two dimensions in a three-dimensional space, and this gives too limited an idea of the possibilities of a four-dimensional surface in space of five or more dimensions. In two dimensions there is just one total curvature and if that vanishes the surface is flat or at least can be unrolled into a plane.

....

Wherever there is matter there is action and therefore curvature ; and it is interesting to notice that in ordinary matter the curvature of the space-time world is by no means insignificant. For example, in water of ordinary density the curvature is the same as that of space in the form of a sphere of radius 570,000,000 kilometres. The result is even more surprising if expressed in time units ; the radius is about half an hour.

It is difficult to picture quite what this means ; but at least we can predict that a globe of water of 570,000,000 km. radius would have extraordinary properties. Presumably there must be an upper limit

[1] Action is determined as energy multiplied by time (Chwolson).

to the possible size of a globe of water. So far as I can make out a homogeneous mass of water of about this size (and no larger) could exist. It would have no centre, and no boundary, every point of it being in the same position with respect to the whole mass as every other point of it—like points on the *surface* of a sphere with respect to the surface. Any ray of light after travelling for an hour or two would come back to the starting point. Nothing could enter or leave the mass, because there is no boundary to enter or leave by; in fact it is co-extensive with space. There could be no other world anywhere else, because there isn't an " anywhere else " (pp. 147, 148).

An exposition of the theories of new physics which stand apart from " relativity " would take too much space. The study of the structure of light and electricity, the study of the atom (the theories of Bohr), and especially the study of the electron (the quantum theory), lead physics along entirely right lines, and if physics really succeeded in freeing itself from the above-mentioned impediments, which arrest its progress, and also from unnecessarily paradoxical theories of relativism, it would some day discover that it knows much more about the true nature of things than might be supposed.

OLD PHYSICS

Geometrical conception of space, that is, consideration of space apart from time. Conception of space as emptiness in which there may or may not be " bodies ".

One time for all that exists. Time measurable on one scale.

Aristotle's principle of the constancy and unity of laws in the whole universe, and, as deduction from this principle, confidence in immutability of recognised phenomena.

Elementary understanding of measure, measurability and incommensurability. Measures taken for everything from outside.

Recognition of a whole series of concepts, difficult to define, such as time, velocity, etc., as primary concepts requiring no definition.

Law of gravitation or attraction and extension of this law to phenomena of falling (weight).

" Universe of flying balls ", both in celestial space and inside atom.

Theories of vibrations, undulatory movements, etc.

Tendency to interpret all phenomena of radiant energy by undulatory vibrations.

Necessity of hypothesis of " æther " in some form or another. Æther as substance of greatest density, and " æther " as substance of greatest rarity.

NEW PHYSICS

Attempts to escape from three-dimensional space by means of mathematics and metageometry. Four coordinates.

Study of the structure of matter and of radiant energy. Study of the atom. Discovery of electrons.

Recognition of velocity of light as limiting velocity. Velocity of light as universal constant.

Definition of fourth coordinate in connection with velocity of light. Time as imaginary quantity. Minkovsky. Recognition of necessity for taking time together with space. Space-time four-dimensional continuum.

New ideas in mechanics. Recognition of possible incorrectness of principle of conservation of energy. Recognition of possible transformation of matter into energy and vice versa.

Attempts to build systems of absolute units of measure.

Establishment of fact of weight of light and of materiality of electricity.

Principle of increase of energy and mass of body in motion.

Special and general principles of relativity; and the idea of necessity for *finite* space in connection with laws of gravitation and distribution of matter in the universe.

Curvature of space-time continuum. Unlimited, but finite universe, measurements of which are determined by density of matter which constitutes it. Spherical or elliptical space.

" Elastic " space.

New theories of structure of atom. Study of electron. Quantum theory. Study of structure of radiant energy.

II

Insufficiency of four coordinates for the construction of a model of the universe
—Absence of approaches to the problem from mathematics—Artificiality of designating
dimensions by powers—Necessary limitation of the universe in relation to dimensions
—Three-dimensionality of motion—Time as a spiral—Three dimensions of time—Six-
dimensional space—" Period of six dimensions "—Two intersecting triangles or a six-
pointed star—Solid of time—" Historical time " as the fourth dimension—Fifth dimension—The " woof " and the " warp "—Limited number of possibilities in every moment
—The Eternal Now—Actualisation of all possibilities—Straight lines—Limitedness
of the infinite universe—The zero dimension—The line of impossibilities—The seventh
dimension—Motion—Four kinds of motion—Division of velocities—Perception of
the third dimension by animals—Velocity as an angle—Limiting velocity—Space—
Heterogeneity of space—Dependence of dimensions on size—Variability of space
—Materiality and its degrees—The world inside the molecule—" Attraction "—
Mass—Celestial space—Emptiness and fullness of space—Traces of motion—Gradations in the structure of matter—Impossibility of describing matter as consisting of atoms
or electrons—The world of interconnected spirals—The principle of symmetry—Infinity
—Infinity in mathematics and in geometry—Incommensurability—Different meanings
of mathematical, geometrical and physical infinity—Function and size—Transition of
space phenomena into time phenomena—Motion passing into extension—Zero quantities
and negative quantities—Inter-atomic extensions—Analysis of the ray of light—Quanta
of light—The electron—Theory of vibrations and theory of emissions—Duration of the
existence of small units—Duration of existence of electron.

Now having examined the principal features of both the " old "
and the " new " physics, we may ask ourselves whether, on the basis
of the material we possess, it is possible to predict the direction which
the future development of physical knowledge will take, and whether
it is possible to build from these predictions a model of the universe,
the separate parts of which will not contradict and mutually destroy
one another. The answer will be that it would not be difficult to
build such a model, or at any rate it would be quite possible, if we
had at our disposal all the necessary measurements of the universe
accessible to us. A new question arises : " Have we all the necessary
measurements ? " And the answer must undoubtedly be : " No, we
have not." Our measurements of the universe are inadequate and
incomplete. In a " geometrical " three-dimensional universe this is
quite clear ; the world cannot be fitted into the space of three coordinates. Too many things are left out, things which cannot be
measured. It is equally clear also in the " metageometrical " universe
of four coordinates. The world with all its variety of phenomena
does not fit into four-dimensional space, no matter how we take the
fourth coordinate, whether as a quantity analogous to the first three
or as an imaginary quantity determinable relatively to the ultimate
physical velocity that has been found, i.e. the velocity of light.

The proof of the artificiality of the four-dimensional world in new physics lies first of all in the extreme complexity of its construction, which requires a *curved space*. It is quite clear that this *curvature* of space indicates the presence in it of yet another dimension or dimensions.

The universe of four coordinates is as unsatisfactory as the universe of three coordinates. And to be more exact we can say that we do not possess all the measurements necessary for the construction of a model of the universe, because neither the three coordinates of old physics nor the four coordinates of new physics are sufficient for the description of all the variety of phenomena in the universe; or, in other words, because we have not enough dimensions.

Let us imagine that somebody builds a model of a house, having only the floor, one wall and the roof. This will be a model corresponding to a *three-dimensional* model of the universe. It will give a general impression of the house, but only on condition that both the model itself and the observer remain motionless. The slightest movement will destroy the whole illusion.

The *four-dimensional* model of the universe of new physics is the same model, only arranged so that it rotates, turning its front always to the observer. This can prolong the illusion for some time, but only on the condition of there being not more than *one observer*. Two people observing such a model from different sides will very soon see in what the trick consists.

Before attempting to make clear without any analogies what it actually means to say that the universe does not fit into three-dimensional or four-dimensional space, and before attempting to discover what number of coordinates really determines the universe, I must eliminate one of the most essential misunderstandings which exists with regard to dimensions.

The is to say, I must repeat that there is no approach from mathematics to the study of the dimensions of space or space-time. And mathematicians who assert that the whole problem of the fourth dimension in philosophy, in psychology, in mysticism, etc., has arisen because "someone once overheard a conversation between two mathematicians on subjects they alone could understand," are greatly mistaken, whether voluntarily or involuntarily is best known to themselves.

Mathematics detaches itself easily and simply from three-dimensional physics and Euclidean geometry, because really it does not belong there at all.

It is quite wrong to think that all mathematical relations must have physical or geometrical meanings. On the contrary, only a very small and the most elementary part of mathematics has a permanent connection with geometry and with physics, and only very few geometrical and physical quantities can have permanent mathematical expression.

For us it is necessary to understand exactly that dimensions cannot be expressed mathematically and that consequently mathematics cannot serve as an instrument for the investigation of problems of space and time. Only measurements along previously agreed-upon coordinates can be expressed mathematically. It can, for instance, be said that the length of an object is 5 metres, the breadth 10 metres and the height 15 metres. But the difference between the *length*, the *breadth* and the *height* themselves cannot be expressed ; mathematically they are equivalent. Mathematics *does not feel* dimensions as geometry and physics feel them. Mathematics cannot feel the difference between a point, a line, a surface and a solid. The point, the line, the surface and the solid can be expressed mathematically only by means of *powers*, that is to say, simply for the sake of designation : a, a line ; a^2, a surface ; a^3, a solid. But the fact is that the same designations would serve also for segments of a line of different lengths :—a, 10 metres ; a^2, 100 metres ; a^3, 1000 metres.

The artificiality of designating dimensions by powers becomes perfectly clear if we reason in the following way:

We assume that a is a line, a^2 is a square, a^3 is a cube, a^4 is a body of four dimensions ; a^5 and a^6, as will be seen later, can be explained. But what will a^{25} mean, or a^{125}, or a^{1000} ? Once we allow that dimensions correspond to powers, this will mean that powers *actually* express the dimensions. Consequently the number of dimensions must be the same as the number of powers. This would be an obvious absurdity, as the limitation of the universe in relation to number of dimensions is quite obvious, and no one would seriously assert the possibility of an infinite or even of a large number of dimensions.

Having established this point, we may note once more, though it should be quite clear already, that three coordinates are not sufficient for the description of the universe, for such a universe would contain no motion, or, putting it differently, every observable motion would immediately destroy the universe.

The fourth coordinate takes time into consideration. Space is no longer taken separately. Four-dimensional space-time allows of motion.

But motion by itself is a very complex phenomenon. At the very first approach to motion we meet with an interesting fact. Motion has in itself three clearly expressed dimensions : duration, velocity and " direction ". But this direction does not lie in Euclidean space, as was assumed by old physics ; it is a direction from before to after, which for us never changes and never disappears.

Time is the measure of motion. If we represent time by a line, then the only line which will satisfy all the demands of time will be a *spiral*. A spiral is a " three-dimensional line ", so to speak, that is, a line which requires three coordinates for its construction and designation.

The three-dimensionality of time is completely analogous to the three-dimensionality of space. We do not measure space by *cubes*, we measure it linearly in different directions, and we do exactly the same with time, although in time we can measure only two coordinates out of three, namely the duration and the velocity ; the direction of time for us is not a quantity but an absolute condition. Another difference is that in regard to space we realise that we are dealing with a three-dimensional continuum, whereas in regard to time we do not realise it. But, as has been said already, if we attempt to unite the three coordinates of time into one whole, we shall obtain a spiral.

This explains at once why the " fourth coordinate " is insufficient to describe time. Although it is admitted to be a curved line, its curvature remains undefined. Only three coordinates, or the " three-dimensional line ", that is, the spiral, give an adequate description of time.

The three-dimensionality of time explains many phenomena which have hitherto remained incomprehensible, and makes unnecessary most of the elaborate hypotheses and suppositions which have been indispensa₁ ₂ in the attempts to squeeze the universe into the boundaries of a th. 'e or even four-dimensional continuum.

This also explains the failure of relativism to give a comprehensible form to its explanations. Excessive complexity in any construction is always the result of something having been omitted or wrongly taken at the outset. The cause of the complexity in this case lies in the above-mentioned impossibility of squeezing the universe into the boundaries of a three-dimensional or. four-dimensional continuum. If we try to regard three-dimensional space as two-dimensional and to explain all physical phenomena as occurring on a surface, several further " principles of relativity " will be required.

The three dimensions of time can be regarded as the continuation of the dimensions of space, i.e. as the " fourth ", the " fifth " and the " sixth " dimensions of space. A " six-dimensional " space is undoubtedly a " Euclidean continuum ", but of properties and forms totally incomprehensible to us. The six-dimensional form of a body is inconceivable for us, and if we were able to apprehend it with our senses we should undoubtedly see and feel it as three-dimensional. Three-dimensionality is a function of our senses. Time is the boundary of our senses. Six-dimensional space is reality, the world as it is. This reality we perceive only through the slit of our senses, touch and vision, and define as three-dimensional space, ascribing to it Euclidean properties. Every six-dimensional body becomes for us a three-dimensional body *existing in time*, and the properties of the fifth and the sixth dimensions remain for us imperceptible.

Six dimensions constitute a " period ", beyond which there can be nothing except the repetition of the same period on a different scale. The period of dimensions is limited at one end by the point, and at the other end by infinity of space multiplied by infinity of time, which in ancient symbolism was represented by two intersecting triangles, or a six-pointed star.

Just as in space one dimension, a line, or two dimensions, a surface, cannot exist by themselves and when taken separately are nothing but imaginary figures, while the *solid* exists in reality, so in time only the three-dimensional *solid of time* exists in reality.

In spite of the fact that the counting of dimensions in geometry begins with the line, actually, in the real physical sense, only the material point and the solid are objects which exist. Lines and surfaces are merely features and properties of a solid. They can also be regarded in another way : a line as the path of the motion of a point in space, and a surface as the path of the motion of a line along the direction perpendicular to it (or its rotation).

The same may be applied to the solid of time. In it only the point (the moment) and the solid are real. The *moment* can change, that is, it can contract and disappear or expand and become a solid. The *solid* also can contract and become a point, or can expand and become an infinity.

The number of dimensions can neither be infinite nor very great ; *it cannot be more than six*. The reason for this lies in the property of the sixth dimension which includes in itself *All Possibilities* of the given scale.

In order to understand this it is necessary to examine the content of the three dimensions of time taken in their " space " sense, that is, as the fourth, the fifth and the sixth dimensions of space.

If we take a three-dimensional body as a point, the line of the existence or motion of this point will be a line of the fourth dimension.

Let us take the line of time as we usually conceive it.

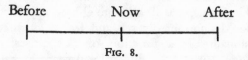

FIG. 8.

The line determined by the three points "before", "now", "after", is a line of the fourth dimension.

Let us imagine several lines perpendicular to this line, before-now-after. These lines, each of which designates *now* for a given moment, will express the perpetual existence of past and possibly of future moments.

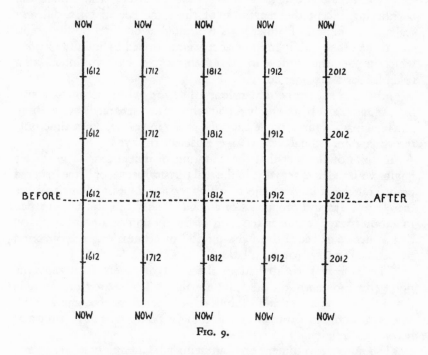

FIG. 9.

Each of these perpendicular lines is the *perpetual now* for some moment, and every moment has such a line of *perpetual now*.

This is the fifth dimension.

The fifth dimension forms a surface in relation to the line of time.

Everything we know, everything we recognise as existing, lies on the line of the fourth dimension; the line of the fourth dimension is the "historical time" of our section of existence. This is the only "time" we know, the only time we feel, the only time we recognise. But though we are not aware of it, sensations of the existence of other "times", both parallel and perpendicular, continually enter into our consciousness. These parallel "times" are completely analogous to our time and consist of before-now-after, whereas the perpendicular "times" consist only of now, and are, as it were, cross-threads, the *woof* in a fabric, in their relation to the parallel lines of time which in this case represent the *warp*.

But each moment of "now" on the line of time, that is, on one of the parallel lines, contains not one, but a certain number of possibilities, at times a great, at others a small number. The number of possibilities contained in every moment must necessarily be limited, for if the number of possibilities were not limited, there would be no impossibilities. Thus each moment of time, within certain limited conditions of being or physical existence, contains a definite number of possibilities and an infinite number of impossibilities. But impossibilities can also be of different kinds. If, walking through a familiar rye-field, I suddenly saw a big birch tree which was not there yesterday, it would be an impossible phenomenon (precisely the "material miracle" which is not admitted by the principle of Aristotle). But if, walking through a rye-field, I saw in the middle of it a coconut palm, this would be an impossible phenomenon of a different kind, also a "material miracle", but of a much higher or more difficult order. This difference between impossibilities should be kept in mind.

On the table before me there are many different things. I may deal with these things in different ways. But I cannot, for instance, take from the table something that is not there. I cannot take from the table an orange that is not there, just as I cannot take from it the Pyramid of Kheops or St. Isaac's Cathedral. It looks as though there was actually no difference in this respect between an orange and a pyramid, and yet there is a difference. An orange *could be* on the table, but a pyramid *could not be*. However elementary all this is, it shows that there are different degrees of impossibility.

But at present we are concerned only with possibilities. As I have already mentioned, each moment contains a definite number of possibilities. I may actualise one of the existing possibilities, that is, I may do something. I may do nothing. But whatever I do, that is, whichever of the possibilities contained in the given moment is actualised, the actualisation of this possibility will determine *the following moment of time*, the following *now*. This second moment of time will again contain a certain number of possibilities, and the actualisation of one of these possibilities will determine *the following moment of time*, the following *now*, and so on.

Thus the line of the direction of time can be defined as the line of the actualisation of one possibility out of the number of possibilities which were contained in the preceding point.

The line of this actualisation will be the line of the fourth dimension, the line of time. We visualise it as a straight line, but it would be more correct to think of it as a zigzag line.

The perpetual existence of this actualisation, the line perpendicular to the line of time, will be the line of the fifth dimension, or the line of eternity.

For the modern mind eternity is an indefinite concept. In ordinary conversational language eternity is taken as a limitless extension of time. But religious and philosophical thought put into the concept of eternity ideas which distinguish it from mere infinite extension homogeneous with finite extension. This is most clearly seen in Indian philosophy with its idea of the *Eternal Now* as the state of Brahma.

In fact, the concept of eternity in relation to time is the same as the concept of a surface in relation to a line. A surface is a quantity incommensurable with a line. Infinity for a line need not necessarily be a line without end ; it may be a surface, that is an infinite number of finite lines.

Eternity can be an infinite number of finite " times ".

It is difficult for us to think of " time " in the plural. Our thought is too much accustomed to the idea of one time, and though in theory the idea of the plurality of " times " is already accepted by new physics, in practice we stil' hink of time as one and the same always and everywhere.

What will the sixth dimension be ?

The sixth dimension will be the line of the actualisation of other possibilities which were contained in the preceding moment but were not actualised in " time ". In every moment and at every point of the three-dimensional world there are a certain number of possibilities ; in " time ", that is, in the fourth dimension, one possibility is actualised every moment, and these actualised possibilities are laid out, one beside another, in the fifth dimension. The line of time, repeated infinitely in eternity, leaves at every point unactualised possibilities. But these possibilities, which have not been actualised in one time, are actualised in the sixth dimension, which is an aggregate of " all times ". The lines of the fifth dimension, which run perpendicular to the line of " time ", form as it were a surface. The lines of the sixth dimension, which start from every point of " time " in all possible directions, form the solid or three-dimensional continuum of time, of which we know only one dimension. We are one-dimensional beings in relation to time. Because of this we do not see parallel time or parallel times ; for the same reason we do not see the angles and turns of time, but see time as a straight line.

Until now we have taken all the lines of the fourth, the fifth and the sixth dimensions as straight lines, as coordinates. But we must remember that these straight lines cannot be regarded as really existing. They are merely an imaginary system of coordinates for determining the spiral.

Generally speaking, it is impossible to establish and prove the real existence of straight lines beyond a certain definite scale and outside certain definite conditions. And even these " conditional straight lines " cease to be straight if we imagine them on a revolving body which possesses, besides, a whole series of other movements. This is quite clear as regards space lines : straight lines are nothing but imaginary coordinates which serve to measure the length, the breadth and the depth of spirals. But time lines are geometrically in no way different from space lines. The only difference lies in the fact that in space we know three dimensions and are able to establish the *spiral* character of all cosmic movements, that is, movements which we take on a sufficiently large scale. But we dare not do this as regards " time ". We try to lay out the whole space of time on one line of the great time which is general for everybody and everything. But this is an illusion ; general time does not exist, and each separately existing body, each separately existing " system " (or what is accepted as such), has its own time. This is recognised by new physics. But what it means and what a separate existence means is not explained by new physics.

Separate time is always a completed circle. We can think of time as a straight line only on the great straight line of the great time. If the great time does not exist, every separate time can only be a circle, that is, a closed curve. But a circle or any closed curve requires two coordinates for its definition. The circle (circumference) is a two-dimensional figure. If the second dimension of time is eternity, this means that eternity enters into every circle of time and into every moment of the circle of time. Eternity is the curvature of time. Eternity is also movement, an *eternal movement*. And if we imagine time as a circle or as any other closed curve, *eternity* will signify eternal movement along this curve, eternal repetition, eternal recurrence.

The fifth dimension is movement in the circle, repetition, recurrence. The sixth dimension is the way out of the circle. If we imagine that one end of the curve rises from the surface, we visualise the third dimension of time—the sixth dimension of space. The line of time becomes a spiral. But the spiral, of which I have spoken

before, is only a very feeble approximation to the spiral of time, only
its possible geometrical representation. The actual spiral of time is
not analogous to any of the lines we know, for it branches off at every
point. And as there can be many possibilities in every moment, so
there can be many branches at every point. Our mind refuses not
only to visualise, but even to think of the resulting figure in curved
lines, and we should lose the direction of our thought in this impasse
if *straight lines* did not come to our aid.

In this connection we can understand the meaning and purpose
of the straight lines of the system of coordinates. Straight lines are
not a naïveté of Euclid, as non-Euclidean geometry and the " new
physics " connected with it are trying to make out. Straight lines
are a concession to the weakness of our thinking apparatus, a concession thanks to which we are able to think of reality in approximate
forms.

A figure of three-dimensional time will appear to us in the form
of a complicated structure consisting of radii diverging from every
moment of time, each of them bearing within it its own time and
throwing out new radii at every point. Taken together these radii
will form the three-dimensional continuum of time.

We live and think and exist on one of the lines of time. But the
second and third dimensions of time, that is, the surface on which
this line lies and the solid in which this surface is included, enter every
moment into our life and into our consciousness, and influence our
" time ". When we begin to feel the three dimensions of time we
call them direction, duration and velocity. But if we wish to understand the true interrelation of things even approximately, we must
bear in mind the fact that direction, duration and velocity are not
real dimensions, but merely the reflections of the real dimensions in
our consciousness.

In thinking of the *time solid* formed by the lines of all the possibilities included in each moment, we must remember that beyond these
there can be nothing.

This is the point at which we can understand the *limitedness of
the infinite universe.*

As has been said before, the three dimensions of space plus the
zero dimension and plus the three dimensions of time form the *period
of dimensions.* It is necessary to understand the properties of this
period. It includes both space and time. The period of dimensions
may be taken as *space-time*, that is, the space of six dimensions or the
space of the actualisation of all possibilities. Outside this space we
can think only of repetitions of the period of dimensions either on
the scale of zero or on the scale of infinity. But these are different

spaces, which have nothing in common with the space of six dimensions and may or may not exist, without changing anything in the space of six dimensions.

The counting of dimensions in geometry begins with the line, the first dimension, and in a certain sense this is right. But both space and time have yet another, the *zero dimension*—the point or the moment. And it must be understood that any space solid, up to the *infinite sphere* of old physics, is a *point* or a *moment* when taken in time.

The zero dimension, the first, the second, the third, the fourth, the fifth and the sixth dimensions form the period of dimensions. But a " figure " of the zero dimension, a point, is *a solid of another scale*. A figure of the first dimension, a line, is infinity in relation to a point. For itself a line is a solid, but a solid of another scale than a point. For a surface, that is, for a figure of two dimensions, a line is a *point*. A surface is three-dimensional for itself, whereas for a solid it becomes a point, and so on. A line and a surface are for us only geometrical concepts, and it is incomprehensible at the first glance how they can be three-dimensional bodies for themselves. But it becomes more comprehensible if we begin with the solid which represents a really existent physical body. We know that a body is three-dimensional for itself as well as for other three-dimensional bodies of a scale near its own. It is also infinity for a surface, which is *zero* in relation to it, because no number of surfaces will make a solid. And the solid is also a point, a zero, a figure of the zero dimension, for the fourth dimension, first, because, however big it may be, a solid is a point, that is, a moment for time, and, second, because no number of solids will make time. The whole of three-dimensional space is but a moment in time. It should be understood that " lines " and " surfaces " are only names which we give to dimensions which for us lie between the point and the solid. They have no real existence for us. Our universe consists only of points and solids. A point is zero dimension, a solid is three dimensions. On another scale a solid must be taken as a time point, and on yet another scale again as a solid, but as a solid of three dimensions of *time*.

In such a simplified universe there would be no time and no motion. Time and motion are created precisely by these *incompletely perceived solids*, that is, by space and time lines and space and time surfaces. And the period of dimensions of the real universe actually consists of *seven powers of solids* (a power is of course only a name in this case). (1) A point,—the hidden solid. (2) A line,—the solid of the second power. (3) A surface,—the solid of the third power. (4) A body or

a solid,—the solid of the fourth power. (5) Time, or the existence of
a body or a solid in time,—the solid of the fifth power. (6) Eternity,
or the existence of time,—the solid of the sixth power. (7) That for
which we have no name, the " six-pointed star ", or the existence of
eternity,—the solid of the seventh power.

Further it should be observed that dimensions are movable, i.e.
any three consecutive dimensions form either " time " or " space ",
and the " period " can move upwards and downwards when one
degree is added above and one is taken away from below or when
one degree is added below and one is taken away from above.   Thus,
if one dimension from " below " is added to the six dimensions we
possess, then one dimension from " above " must disappear.   The
difficulty of understanding this eternally changing universe, which
contracts and expands according to the *size of the observer* and the
speed of his perception, is counterbalanced by the constancy of
laws and relative positions in these changing conditions.

The " seventh dimension " is impossible, for it would be a line
leading nowhere, running in a non-existent direction.

The line of impossibilities is the line of the seventh, the eighth and
the other non-existent dimensions, a line which leads nowhere and
comes from nowhere.   No matter what strange universe we may
imagine, we can never admit the real existence of a solar system in
which the moon is made of green cheese.   In the same way, whatever strange scientific manipulations we may think of, we cannot
imagine that Prof. Einstein would really erect a pole on the Potsdamer
Platz in order to measure the distance between the earth and the
clouds, as he threatens to do in his book.

One could find many such examples.   The whole of our life
actually consists of phenomena of the " seventh dimension ", that is,
of phenomena of fictitious possibility, fictitious importance and fictitious value.   We live in the seventh dimension and cannot escape
from it.   And our model of the universe can never be complete if
we do not realise the place occupied in it by the " seventh dimension ".
But it is very difficult to realise this.   We never even come near to
understanding how many *non-existent* things play a rôle in our life,
govern our fate and our actions.   But again, as has been said before,
even the non-existent and the impossible can be of different degrees—
and therefore it is perfectly justifiable to speak not of the seventh
dimension, but generally of *imaginary dimensions*, the number of which
is also imaginary.

In order to establish with complete exactitude the necessity for
regarding the world as a world of six coordinates, it is necessary to

examine the fundamental concepts of physics, which have remained without definition, and see whether it is not possible to find definitions for them with the help of some of the principles we have established above.

We will deal with matter, space, motion, velocity, infinity, mass, light, etc.

We will begin with motion.

In the usual views of both the old and the new physics motion remains always the same. Distinction is made only between its properties : duration, velocity, direction in space, discontinuity, continuity, periodicity, acceleration, retardation and so on, and the characteristics of these properties are attributed to motion itself, so that motion is divided into rectilinear, curvilinear, continuous, non-continuous, accelerated, retarded, etc. The principle of the relativity of motion led to the principle of the composition of velocities, and the working out of the principle of relativity led to the denial of the possibility of the composition of velocities when " terrestrial " velocities are compared with the velocity of light. This led to many other conclusions, suppositions and hypotheses. But these do not interest us for the moment. One fact, however, must be established, namely, that the very concept " motion " is not defined. Equally " velocity " is not defined. In regard to " light ", opinions of physicists diverge.

For the present it is only important for us to realise that motion is always taken as a phenomenon of one kind. There are no attempts to establish different kinds of phenomena in motion itself. And this is especially strange, because for direct observation there definitely exist four kinds of motion as four perfectly distinct phenomena.

In certain cases direct observation deceives us, for instance when it shows much non-existent motion. But phenomena themselves are one thing, and the division of them is another. In this particular case direct observation brings us to real and unquestionable facts. One cannot reason about motion without having understood the division of motion into four kinds.

These four kinds of motion are as follows :

1. *Slow motion, invisible as motion, for instance the movement of the hour-hand of a clock.*

2. *Visible motion.*

3. *Quick motion, when a point becomes a line, for instance the movement of a smouldering match waved quickly in the dark.*

4. *Motion so quick that it does not leave any visual impression, but produces definite physical effects, for instance the motion of a flying bullet.*

In order to understand the difference between the four kinds of motion let us imagine a simple experiment. Let us imagine that we are looking at a white wall at a certain distance from us on which a black point is moving, now faster, now slower, then stopping altogether.

It is possible to determine exactly when we begin to see the point move and when we cease to see it move.

We see the movement of the point *as movement* if the point covers in $\frac{1}{10}$th of a second one or two minutes of the arc of a circle, taking as the radius our distance from the wall. If the point moves more slowly it will appear to us motionless.

Let us suppose first that the point moves with the velocity of the hour-hand of a clock. Comparing its position with other, motionless, points, first, we establish the fact of the movement of the point and, second, we determine the velocity of its movement; but we do not see the movement itself.

This will be the first kind of motion, *invisible motion.*

Further, if the point moves more quickly, covering two minutes of arc or more in $\frac{1}{10}$th of a second, we see its motion as motion.

This is the second kind of motion, *visible motion.* It can be very varied in its character and cover a large scale of velocities, but when velocity is increased 4,000 to 5,000 times, and in certain cases less, it passes into the *third kind of motion.*

This means that if the point moves very fast, covering in $\frac{1}{10}$th of a second the whole field of our vision, i.e. 160° or 9,600 minutes of arc, we shall see it not as a moving point but *as a line.*

This is the third kind of motion, with a visible trace, or motion in which the moving point is transformed into a line, motion with the apparent addition of one dimension.

And, finally, if the point starts off at once with the velocity of, say, a rifle bullet, we shall not see it at all, but if the " point " possesses weight and mass, its motion may have many physical effects which we can observe and study. For instance we can hear the motion, we can see other motions aroused by the invisible motion, and so on.

This is the fourth kind of motion, motion with an invisible but perceptible trace.

These four kinds of motion are absolutely real facts upon which depend the whole form, aspect and correlation of phenomena in our universe. This is so because the distinction of the four kinds of motion is not only subjective, i.e. they differ not only in our perception, but they *differ physically* in their results and in their action on other phenomena ; and above all they are different in relation to one another, and this relation is permanent.

The ideas that have been set forth here may appear very naïve to a learned physicist.—What is the eye ? he would say. The eye has a strange capacity for " remembering " for about $\frac{1}{10}$th of a second what it has seen ; if the point moves sufficiently fast for the *memory* of each $\frac{1}{10}$th of a second to merge with another *memory*, the result will be a line. There is no transformation of a point into a line here. It is all entirely subjective, that is, it all takes place only in us, only in our perception. In reality a moving point remains a moving point.

This is how the matter appears from the scientific point of view.

The objection is based on the supposition that we *know* that the observed phenomenon is produced by the motion of a point. But suppose we do not know ? How can we ascertain it if we cannot come sufficiently near the line we observe, or arrest the motion, stop the supposed moving point ?

Our eye sees a line ; with a certain velocity of motion, a photographic camera will also " see " a line or a streak. The moving point is actually transformed into a line. We are quite wrong in not trusting our eye in this case. This is just a case in which our eye does not deceive us. The eye establishes an exact principle of division of velocities. The eye certainly establishes these divisions for itself, on its own level, on its own scale. And this scale may change. What will not change, for instance in connection with the distance, what will remain the same on any scale, is, first, the number of different kinds of motion—there will always be four—and next, the *interrelation* of the four velocities with their derivatives, i.e. with their results, or the interrelation of the four kinds of motion. This interrelation between the four kinds of motion creates the whole visible world. And the essence of this interrelation consists in the fact that one motion is not necessarily motion relatively to another motion, but only if the velocities which are compared do not differ greatly from one another.

Thus in the above example the visible motion of the point on the wall is *motion* in comparison both with invisible motion and with motion fast enough to form a line. But it will not be motion in relation to a flying bullet, for which it will be immobility, just as the line formed by a swiftly moving point will be a line and not motion for a slowly (invisibly) moving point. This can be formulated in the following way :

Dividing motion into four kinds, according to the above principles, we observe that motion is motion (with increased or decreased velocity) only for kinds of motion that are near one another, that is, within the limits of a definite correlation of velocities, or, to put it more precisely, within the limits of a certain definite increase or decrease of velocity, which can probably be determined exactly. More remote /

kinds of motion, i.e. motions with very different velocities, for instance, 4,000 or 5,000 times slower or faster than another, are for one another not motions of different velocity, but phenomena of a greater or lesser number of dimensions.

But what is velocity? What is this mysterious property of motion which exists only in middle degrees and disappears in small and large degrees, thus subtracting or adding one dimension? And what is motion itself?

Motion is an apparent phenomenon dependent upon the extension of a body in the three dimensions of time. This means that every three-dimensional body possesses also three time-dimensions which we do not see as such and which we call the properties of motion or of existence. Our mind cannot embrace time-dimensions in their entirety, there exist no concepts which would express their essence in all their variety, for all existent " time concepts " express only one side, or only one dimension, each. Therefore the extension of three-dimensional bodies in the indefinable (for us) three dimensions of time appears to us as motion with all its properties.

We stand in exactly the same position in relation to dimensions of time as animals stand in relation to the third dimension of space.

I wrote in *Tertium Organum* about the perception of the third dimension by animals. All apparent movements are real for them. A house turns about when a horse runs past it, a tree jumps into the road. Even if an animal is motionless and only examines an equally motionless object, this object begins to manifest strange movements. The animal's own body, even in the state of rest, may manifest for it many strange movements, which our bodies do not manifest for us.

Our relation to motion and especially to velocity is very similar to this. Velocity can be a property of space. The sensation of a velocity may be the sensation of the penetration into our consciousness of one of the dimensions of a higher space unknown to us.

Velocity can be regarded as an *angle*. And this at once explains all the properties of velocity and especially the fact that both great and small velocities cease to be velocities. An angle has naturally a limit both in one direction and in the other.

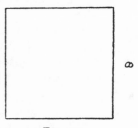

FIG. 10.

Let us again imagine a world of flat beings. Let us imagine these flat beings in the shape of squares with their organs of perception situated on one side of the square. Let us call this percipient side *a*.

Let us imagine that the "square" is turned with its percipient side towards two figures, let us say two "triangles" ABC and DEF, in the position shown in the diagram.

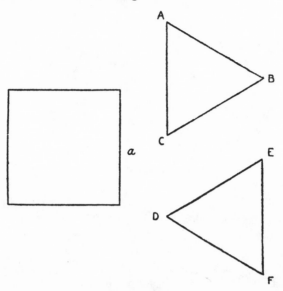

FIG. 11.

Of the triangle ABC it knows only the line AC, and this line is motionless for it. Of the triangle DEF it knows the lines DE and DF, which appear to it as one line, and these lines, which go out of the field of its vision, must undoubtedly differ from the line AC, possess some property which the line AC does not possess. The "square" will call this property *motion*.

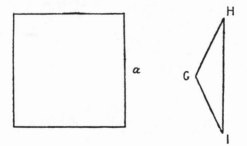

FIG. 12.

If the "square" happens to meet the triangle GHI, the lines GH and GI will also be "motion" for it, but a slower motion.

And if the " square " meets the triangle JKL, the lines JK and JL will be a swifter motion.

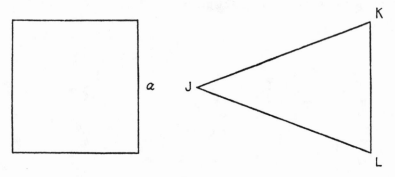

FIG. 13.

And finally, if the " square " meets lines almost perpendicular to its percipient side, like the lines MN and MO, it will say that this is the limiting, maximal velocity and that there can be no higher velocity.

The idea of velocity as an angle makes not only clear but necessary the idea of a limiting velocity beyond which there can exist no other velocity, and also the idea of the impossibility of an infinite velocity, because an angle cannot be infinite and must have a limit which can always be established and measured.

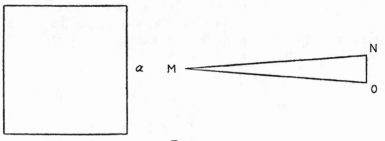

FIG. 14.

So far, in all the above examples velocity has been taken as uniform and unchangeable. But, on the basis of the same principle, it is easy to establish the meaning of acceleration, variable velocity, and so on.

Let us imagine that the receding line PQ is not a straight line but a line with an angle.

The flat being in examining such a line from the point P will see this line as motion starting with one speed and then accelerating.

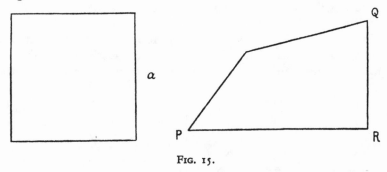

FIG. 15.

The line ST will appear to it as a motion alternately accelerated and retarded. And further, lines with angles, curves of different kinds, lines lying at various or changing angles to the percipient side, will represent different kinds of velocity : constant, variable, uniformly accelerated, uniformly retarded, periodically accelerated and retarded and so on.

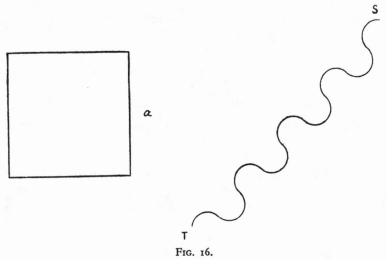

FIG. 16.

The essence of all that has been said is that a line receding at an angle will appear as motion only if it lies at angles of certain definite degrees. A line lying at a very small angle to a motionless line which is parallel to the percipient side would appear motionless ; at a greater angle it would appear as motion, and a line lying at an angle approaching the limit would appear something altogether different from motion. Thus "velocity" is only the property of certain definite angles, and as the angle does not depend on scale, it is quite possible that "velocity" is the only constant phenomenon in the universe.

This principle is in no way changed by the alteration of the angles on a spherical surface, or for instance on the saddle-shaped surface used by Lobatchevsky, in comparison with the angles on a

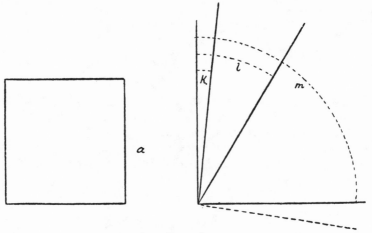

Fig. 17.

Angle *k*—small velocity, the beginning of motion.
Angle *l*—greater velocity, visible motion.
Angle *m*—limiting velocity, the end of motion.
Dotted line below—an impossible acceleration.

flat surface, because for every kind of surface the angles will remain unchangeable.

Now, starting from the above definitions of time, motion and velocity, we shall pass to the definition of space, matter, mass, gravitation, infinity, commensurability and incommensurability, "negative quantity", etc.

As regards space, the first fact we come upon is that space is much too readily accepted as *homogeneous*. The very question of the possible heterogeneity of space never arises. And if such a question ever arose, it was only in the domain of purely mathematical speculation and never passed into conceptions of the real world from the point of view of heterogeneous space.

Even the most complex mathematical and metageometrical views assert themselves each to the exclusion of all the others. "Spherical" space, "elliptical" space, space determined by the density of matter and by the laws of gravitation, " finite and yet limitless " space—in each case this is *the whole* of space, and in each case the whole of space is uniform and homogeneous.[1]

Of all the latest definitions of space the most interesting is the " mollusc " of Einstein. The " mollusc " anticipates many future discoveries. The " mollusc " is able to move by itself, to expand and to contract. The " mollusc " can be unequal to itself and heterogeneous with itself.

But still the " mollusc " is only an analogy, only a very timid example of the way in which space can and should be regarded. And behind this example, in order to make it possible, the whole arsenal of mathematics, metageometry and new physics with the " special " and " general " principles of relativity is necessary.

In reality all this could be done much more simply, if only the possible heterogeneity of space were understood.

Let us take space just as we took motion, from the point of view of direct observation.

(A) The space, occupied by the house in which I live, by the room in which I am now and by my body, is perceived by me as three-dimensional. Certainly this is not a pure " percept ", for it has already passed through the prism of thinking, but as the three-dimensionality of the house, the room and my body does not give rise to argument, it can be accepted.

(B) I look out of the window and see a portion of the sky with several stars in it. *The sky is two-dimensional for me.* My mind knows that the sky possesses " depth ". But my direct senses do not tell me so. On the contrary, they deny the truth of it.

(C) I am reflecting on the structure of matter and on a unit such as a molecule. One molecule has no dimensions for the direct senses but, by reasoning, I come to the conclusion that the space occupied

[1] The present chapter in its essential features was completed in 1912. The first part was written later, but in making a survey of the present state of physics I did not try to bring it fully up to date and to mention *all* the theories that had appeared by that time, because not one of them changed anything in my principal conclusions. The most complete exposition of views on space will be found by the reader in Prof. Eddington's book, *Space, Time and Gravitation*, particularly in the chapter, " Kinds of Space ". At the beginning of this chapter Prof. Eddington quotes W. K. Clifford (1845–1879) who wrote in his book, *Common Sense of the Exact Sciences* :

" The danger of asserting dogmatically that an axiom based on the experience of a limited region holds universally will now be to some extent apparent to the reader. It may lead us to entirely overlook, or when suggested at once reject, a possible explanation of phenomena. The hypothesis that space is not flat, and again that its geometrical character may change with the time, may or may not be destined to play a great part in the physics of the future ; yet we cannot refuse to consider them as possible explanations of physical phenomena, because they may be opposed to the popular dogmatic belief in the universality of certain geometrical axioms—a belief which has risen from centuries of indiscriminating worship of the genius of Euclid.

This may have a connection with the idea of the heterogeneity of space.

by the molecule, consisting of atoms and electrons, must have six dimensions : three space-dimensions and three time-dimensions, for otherwise, if the molecule did not possess the three time-dimensions, its three space-dimensions would be unable to produce any impression on my senses. A great quantity of molecules produces on me the impression of *matter* possessing mass only because of the six-dimensionality of the space occupied by every molecule.

Thus " space " is not homogeneous for me. The room is three-dimensional, the sky two-dimensional. The molecule has no dimension for direct perception ; atoms and electrons have still less dimension, but owing to their six-dimensionality a multitude of molecules produces on me the impression of matter. If the molecules had no time-dimensions *matter* would be emptiness for me.

What has been said above must leave several points requiring explanation. First, if the molecule has *no dimension* how can atoms and electrons have *still less* ? And second, how do time-dimensions affect our senses and why would not space-dimensions by themselves produce any effect on us ?

In order to answer these questions it is necessary to enlarge upon the above considerations.

A star which appears to me as a twinkling point actually consists of two enormous suns each surrounded by a whole series of planets and separated by colossal distances. This twinkling point in reality occupies an enormous expanse of three-dimensional space.

Here again the objection may be raised, just as in the case of the four kinds of motion, that I take purely subjective sensations and attribute to them real meaning. And again, as in the case of the four kinds of motion, I may reply to this that what interests me is not sensations, but the interrelations of their *causes*. The causes are not subjective, but depend upon perfectly definite and perfectly objective conditions, namely, comparative magnitude and distance.

The house and the room are three-dimensional for me, by virtue of their commensurability with my body. The " sky " is two-dimensional, because it is remote. The " star " is a point because it is small as compared with the " sky ". The " molecule " may be six-dimensional, but as a point, i.e. taken as a zero-dimensional body, it cannot produce any effect on my senses. These are all facts, there is nothing subjective in them.

But this is by no means all.

The dimensions of my space depend upon the size of my body. If the size of my body could change, the dimensions of the space around me would change also. " Dimension " corresponds to " size ".

If the dimensions of my world can change with a change in my size, then the size of my world also can change.

But in what respect?

A right answer to this question will at once put us on the right road.

The smaller the "reference-body" or "reference-system", the smaller the world. Space is proportionate to the size of the reference-body, and all measurements of space are proportionate to the measurements of the reference-body. And yet it is the same space. Let us take an electron on the sun in its relation to visible space and to the earth. For the electron the whole of visible space will be (of course only approximately) a sphere one kilometre in diameter; the distance from the sun to the earth will be a few centimetres, and the earth itself will be almost a "material point". A ray of light from the sun reaches the earth (for the electron) instantaneously. This explains why we can never intercept a ray of light half-way.

If instead of an electron we take the earth, then for the earth distances will necessarily be much longer than they are for us. They will be longer by exactly as many times as the earth is bigger than the human body. This is necessarily so if only because otherwise the earth could not feel itself the three-dimensional body we know it to be, but would be for itself some incomprehensible six-dimensional continuum. But such a self-feeling would contradict the rightly understood principle of the unity of laws. The reason is that if the earth could be for itself a six-dimensional continuum, then we also should have to be for ourselves six-dimensional continua. And since we are for ourselves three-dimensional bodies, the earth also must be for itself a three-dimensional body; although at the same time it is not possible to assert with certainty that the earth's conception of itself must necessarily coincide with our conception of it.

If we now try to imagine what the space occupied by terrestrial objects must be for the electron on the one hand and for the earth on the other, we shall come to a very strange and at first glance paradoxical conclusion. Things which surround us, tables, chairs, objects of daily use, other people, etc., cannot exist for the earth, for they are too small for it. It is impossible to conceive a chair in the planetary world. It is impossible to conceive an individual man in relation to the earth. An individual man cannot exist in relation to the earth. The whole of humanity cannot exist by itself in relation to the earth. It exists only together with all the vegetable and animal world and with all that has been made by the hand of man.

There can be no serious objection to this, because a particle of matter that is as small in relation to the human body as the human body, or even all humanity, is in relation to the earth certainly cannot exist for us. And it is quite obvious that a chair cannot exist in the planetary world because it is too small. What is strange and what is paradoxical is the inference that a chair cannot exist *for the electron or in the world of electrons also, and also because it is too small.*

This seems an absurdity. " Logically " it ought to be that a chair cannot exist for the electron, because a chair is too big compared with the electron. But it would be so only in a " logical ", that is, in a three-dimensional, universe with a permanent space. The six-dimensional universe is illogical and the space in it can contract and expand on an incredibly large scale, preserving only one permanent property, namely *angles.* Therefore, the space existing for the electron which is proportionate to its size will be so small that a *chair* will occupy practically no room in this space.

Thus we have come to a space which expands and contracts in accordance with the size of the " reference-body "—an expandable and contractible space. Einstein's " mollusc " is the nearest approximation to this idea in new physics. But like most of the ideas of new physics, the " mollusc " is not so much a formulation of something new as an attempt to show that the old will not do. The " old " in this case is immovable and unchanging space. The same can be said of the general idea of the space-time continuum. New physics recognises that space cannot be examined apart from time, time cannot be examined apart from space, but what actually constitutes the essence of the relation of space to time and why phenomena of space and phenomena of time appear to be different for direct perception, new physics does not state.

The new model of the universe establishes exactly the unity of space and time, and the difference between them ; it establishes also the principle that space can pass into time and time into space.

In old physics space is always space, and time is always time. In the new physics the two categories make one, *space-time.* In the new model of the universe the phenomena of one category can pass into the phenomena of the other category, and vice versa.

When I write of space, space-concepts and space-dimensions, I mean space for us. For the electron, and most probably even for bodies much larger than the electron, our space is time.

The six-pointed star which represented the world in ancient symbolism is in reality the representation of space-time or the " period of dimensions ", i.e. of the three space-dimensions and the three time-dimensions in their perfect union, where every point of space includes the whole of time and every moment of time includes the whole of space ; when *everything* is *everywhere* and *always.*

But this state of six-dimensional space is incomprehensible and inaccessible to us, for our sense-organs and our mind enable us to establish a connection only with the material world, that is, with a world of certain definite limitations in relation to higher space. We can never see a six-pointed star.

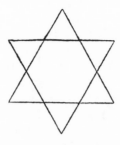

FIG. 18.

What does material world mean? What does materiality mean? What does matter mean?

Earlier in this chapter a definition by Prof. Chwolson was quoted:

> In objectifying the cause of a sensation, that is, transferring this cause into a definite place in space, we conceive this place as containing something which we call *matter* or substance (Vol. I, p. 2).

And further:

> The use of the term "matter" was reserved exclusively for matter which is able to affect our organ of touch more or less directly (Vol. I, p. 6).

Modern physics and chemistry have achieved much in the study of the structure and composition of matter, and they do not limit themselves by a definition of matter like that made by Prof. Chwolson and apparently regard as *matter* everything that admits of objective study, everything that can be measured and weighed, even indirectly. In studying the structure and composition of matter these sciences deal with divisions of matter which are so small that they can produce no effect on our organs of touch, but are nevertheless recognised as material.

In fact both the old view, which limited the concept of matter too closely, and the new view, which extends it too far, are incorrect.

In order to avoid contradictions, indefiniteness and confusion of terms, it is necessary to establish the existence of several *degrees of materiality*.

1. Solid, liquid and gaseous states of matter (up to a certain degree of rarefaction), that is, states in which matter can be divided into " particles ".

2. Very rarefied gases, consisting of separate molecules, and molecules resolved into component atoms.

3. Radiant energy (light, electricity, etc.), that is, the *electronic state of matter*, or electrons with their derivatives not bound into atoms. Certain physicists regard this state as *decomposition of matter*. But there are no data which justify this view.

It is not known how electrons become combined into atoms, just as it is not known how molecules become combined into cells and into protoplasm in living or organised matter.

It is necessary to keep in view these divisions because without applying them it is impossible to find a way out of the chaos in which physical sciences find themselves.

What do these divisions signify from the standpoint of the above principles of " the new model of the universe ", and how can the degrees of materiality be defined ?

Matter of the first kind is three-dimensional, i.e. any part of this matter and any " particle " can be measured in length, breadth and height and exists in time, i.e. in the fourth dimension.

Matter of the second and third kinds, i.e. its components, molecules, atoms and electrons, have no space-dimensions in comparison with particles of matter of the first kind, and reach our consciousness only in large masses and only through their time-dimensions, the fourth, the fifth and the sixth ; in other words, they reach it only by virtue of their motion and the repetition of their motion.

Thus only the first degree of matter can be taken as existing in geometrical forms and in three-dimensional space. Atomic and electronic matter can with every right be regarded as matter belonging not to our, but to another, space, for it requires for its description six dimensions. And its units, molecules, atoms and electrons, if taken by themselves, can with every right be called *immaterial*.

" Materiality " is divided for us into three categories or three degrees.

The first kind of materiality is the state of matter of which our bodies consist. This matter and any part of it must possess (for us) three space-dimensions and one time-dimension ; their fifth and sixth dimensions we cannot perceive.

In the materiality of the first kind there is (for us) more space than time.

The second and the third kinds of materiality are the states of molecules, atoms and electrons, which (for the direct senses) have the zero dimension in space and reach our consciousness by virtue of their three dimensions of time.

In the materiality of the second and third kinds there is (for us) more time than space.

The change of the state of matter from solid to liquid and from liquid to gaseous concerns molecules only, i.e. the distance between them and their cohesion. But inside the molecules, in all three states of matter, the solid, the liquid and the gaseous, everything remains the same, i.e. the proportion of matter and emptiness does not alter. " Electrons " remain equally far one from another inside the atoms and revolve in their orbits in the same way in all states of cohesion of molecules. Changes in the density of matter, i.e. transition from the solid state into the liquid or the gaseous state, do not reach them and do not affect them in any way.

The world inside the molecules is completely analogous to the great space in which celestial bodies move. Electrons, atoms, molecules, planets, solar systems, agglomerations of stars—all these are phenomena of the same order. Electrons move in their orbits in the atom just as planets move in the solar system. Electrons are the same celestial bodies as planets, even their velocity is the same as the velocities of the planets. In the world of electrons and atoms it is possible to observe all the phenomena which are observed in the astronomical world. There are comets in this world which travel from one system to another, there are shooting stars, there are streams of meteorites. " As above so below." Science seems to have proved the old formula of the Hermetists. Unfortunately, however, *it only seems so*, for in actual fact the model of the universe which science builds is too unstable and can fall to pieces at a single touch.

Indeed, what links together all these revolving particles or aggregations of matter ? Why do not the planets of the solar system fly apart in different directions ? Why do they continue to revolve in their orbits round the central luminary ? Why do electrons remain linked with one another, thus constituting an atom ? Why do they not fly apart, why does matter not resolve into nothing ?

Science has always been confronted with these questions in one form or another, and even in our day it is unable to answer them without introducing two new unknown quantities : " attraction " or " gravitation " and " æther ".

" Attraction "—replies science to the above question—keeps the planets near the sun and binds electrons into one whole ; attraction, that mysterious force, the influence of a larger mass upon a smaller mass. This again produces a question : how can one mass influence another, even a smaller one, when it is at a great distance from it ? If we imagine the sun as a large apple, the earth will be a poppy seed at a distance of twelve paces from this apple. How can the apple influence the poppy seed twelve paces from it ? They must be linked in some way, for otherwise the influence of one body upon another remains totally incomprehensible and is in fact impossible.

Scientists have tried to find an answer to this problem by imagining a certain *medium* through which influence is transmitted and in which electrons and (possibly) also celestial bodies revolve.

All these hypotheses, and also the hypothesis of gravitation, are entirely unnecessary, from the point of view of the new model of the universe.

Atomic matter makes our consciousness aware of its existence through its motion. If the motion inside atoms were to stop, matter would turn into emptiness, into nothing. The effect of materiality, the impression of mass, is produced by the *motion* of the minutest particles, which *demands time*. If we take away time, if we imagine atoms without time, that is, if we imagine all the electrons constituting the atom as immovable, *there will be no matter*. *Motionless* small quantities are outside our scale of perception. We perceive not them, but their orbits, or the orbits of their orbits.

Celestial space is emptiness for us, that is, precisely what matter would be without time.

But in the case of celestial space we have learned sooner than we learned in the case of matter that what we see does not correspond to reality, though our science is still far from the right understanding of this reality.

Luminous points have turned into worlds moving in space. The universe of flying globes has come into being. But this picture is not the end of the possible understanding of celestial space.

If we represent schematically the interrelation of celestial bodies, we shall represent them as discs or points at a great distance from one another. But we know that they are not immovable, we know that they revolve round one another, and we know that they are not points. The moon revolves round the earth, the earth revolves round the sun, the sun in its turn revolves round some other luminary unknown to us, or, at any rate, moves in a definite direction along a definite line. Consequently the moon in revolving round the earth at the same time revolves round the sun and at the same time moves somewhere together with the sun. And the earth in revolving round the sun at the same time revolves round an unknown centre.

If we wish to represent graphically the paths of this motion, we shall represent the path of the sun as a line, the path of the earth as a spiral winding round this line, and the path of the moon as a spiral winding round the spiral of the earth. If we wish to represent the path of the whole solar system, we shall have to represent the paths of all the planets and asteroids as spirals winding round the central line of the sun, and the paths of the planets' satellites as spirals round the spirals of the planets. Such a drawing would be very difficult to make, in fact with asteroids it would be impossible ; and it would be still more difficult to construct an exact model from this drawing, especially if all the interrelations, distances, exact thickness of the spirals, etc., were to be strictly observed. But if we were to succeed in building such a model, it would be an exact model of a small particle of *matter* enlarged many times. The same model, reduced a required number of times, would appear to us as impenetrable matter, exactly identical with all the matter which surrounds us.

Matter or substances of which our bodies and all the objects surrounding us consist is built in exactly the same way as the solar system ; only we are incapable of perceiving electrons and atoms as immovable points but perceive them in the form of the complex, and entangled traces of their movement which produce the effect of mass. If we were able to perceive the solar system on a much smaller scale, it would produce on us the effect of matter. There would be no emptiness in the solar system for us, just as there is no emptiness in the matter surrounding us.

The emptiness or fullness of space depends entirely upon the dimensions in which we perceive the matter or particles of matter contained in that space. And the dimensions in which we perceive this matter depend upon the size of the particles of this matter in comparison with our body, upon the greater or lesser distance separating us from them, and upon our perception of their motion (which depends upon the velocity of their own motion and the rate of our perception), which creates the subjective aspect of the world.

All these conditions, taken together, determine the dimensions in which we perceive various agglomerations of matter.

A whole world, consisting of several suns, with their surrounding planets and satellites, rushing with terrific velocity through space, but separated from us by great distances, is perceived by us as an immovable point.

The almost immeasurably small electrons when moving are transformed into lines, and these lines intertwining among themselves create for us the impression of mass, i.e. of hard, impenetrable matter, of which the three-dimensional bodies surrounding us consist.

Matter is created by the fine web made by the traces of the motion of the smallest " material points ".

The study of the principles of this motion is necessary for the understanding of the world, because it is only when we make these principles clear to ourselves that we shall have an exact conception of how the web created by the motion of the electrons is woven and thickened, and how the whole world of infinite variety of phenomena is constructed from this web.

The main principle of the structure of matter from the point of view of the new model of the universe is the idea of *gradations* in this structure. Matter of one kind cannot be described as consisting of units of matter of another kind. It is the greatest mistake to say that tangible matter consists of atoms and electrons. *Atoms* consist of positive and negative electrons. *Molecules* consist of atoms. *Particles of matter* consist of molecules. *Material bodies* consist of matter. It cannot be said that material bodies consist of molecules or of atoms. Atoms and molecules cannot be regarded as material particles ; they belong to a different space-time. It was pointed out earlier that they contain more time than space. An electron is much more a time unit than a space unit.

To regard, for instance, the body of man as consisting of electrons or even of molecules is as wrong as it would be wrong to regard the population of a large town or a company of soldiers or any gathering of people as consisting of *cells*. It is evident that the population of a large and even of a small town, or a company of soldiers, consists not of microscopic cells, but of individual men. Precisely in the same way the body of man consists of individual cells, or simply physically, of matter. Of course I have not in view a metaphor which would regard a gathering of people as an organism and individual people as cells of this organism.

A whole series of unnecessary hypotheses falls away as soon as we realise the general connectedness and cohesion which follow from the above definitions of matter and mass.

The first which falls away is the hypothesis of gravitation. Gravitation is necessary only in the " world of flying balls " ; in the world of *interconnected spirals* it becomes unnecessary. Similarly there disappears the necessity of recognising a " medium " through which gravitation, or " action at a distance ", is transmitted. Everything is connected. The world constitutes One Whole.

Another interesting problem is disclosed at the same time. The hypothesis of gravitation was connected with observations of phenomena of weight and falling. According to the Newtonian legend indeed (the apple which Newton saw fall from the tree), these observations gave grounds for the building up of the whole hypothesis. It occurred to nobody that the phenomena which were explained by " gravitation " or " attraction " on the one hand, and the phenomena of " weight " on the other hand, are *totally different phenomena* having nothing whatever in common.

The sun, the moon, the stars, which we see, are cross-sections of spirals which we do not see. These cross-sections do not fall out of the spirals because of the same principle by reason of which the *cross-section of an apple* cannot fall out of the apple.

But the apple falls to the ground *as though aiming at the centre of the earth* in virtue of an entirely different principle, namely the " *principle of symmetry* ". In Chapter II of this book there is a description of that particular movement which I called movement from the centre and towards the centre along radii, and which, with its laws enumerated there, is the foundation and cause of the phenomena of symmetry.

The laws of symmetry, when they are established and elaborated, will occupy a very important place in the new model of the universe. And it is quite possible that what is called the law of gravitation, in the sense of the formula for calculation, will prove to be a partial expression of the law of symmetry.

The definition of mass as the result of the motion of invisible points dispenses with any necessity for the hypothesis of æther. A ray of light has material structure, and so has electric current ; but light and electricity are matter not formed into atoms, but remaining in the electronic state.

Returning to the concepts of physics and geometry, I must repeat that the wrong development of scientific thought which has led in new physics to the unnecessary complication of problems which were simple in their essence was to a great extent due to operating with *undefined* concepts.

One of these undefined concepts is " infinity ".

Infinity has a definite meaning only in mathematics. In geometry infinity needs to be defined, and still more does it need to be defined in physics. These definitions do not exist, nor have there even been attempts at such definitions that are worthy of attention. " Infinity " is taken merely as something very big, bigger than anything else we can conceive, and at the same time as something completely homogeneous with the finite, yet incalculable. In other words, it is never said anywhere in a definite and exact form that the infinite is *not homogeneous* with the finite. I mean that it has not been established exactly *what* distinguishes the infinite from the finite either physically or geometrically.

In reality, both in the domain of geometry and in the domain of physics, infinity has a distinctive meaning, which differs very greatly from the strictly mathematical meaning. And the establishment of different meanings of infinity solves a number of otherwise insoluble problems and leads our thought out of a series of mazes and blind alleys created either artificially or through misunderstanding.

First of all, an exact definition of infinity dispenses with the necessity for mixing up physics with geometry, which is the favourite idea of Einstein and the foundation of non-Euclidean geometry. I have pointed out earlier that the mixing up of physics and geometry, that is to say, the introduction of physics into geometry, or a physical revaluation of geometrical values (all these rigid rods and non-rigid rods and so on), which follows from an identical mathematical valuation of geometrical and physical values, is unnecessary either for arguments concerning relativity, or for anything else.

Physicists are quite right in feeling that geometry is not sufficient for them; in Euclidean space there is not enough room for them with their luggage. But the remarkable feature of the geometry of Euclid (and this is exactly why Euclidean geometry *should be preserved intact*) consists in the fact that it contains within itself an indication of the way out. There is no need to break up and destroy the geometry of Euclid. It can very well adapt itself to any kind of physical discovery. *And the key to this is infinity*.

The difference between infinity in mathematics and infinity in geometry is quite clear at the first glance. Mathematics does not establish two infinities for one finite quantity. Geometry begins with this.

Let us take any finite line. What is infinity for this line ? We have two answers : a line continued into infinity, or the square, of which the given line is a side. What is infinity for a square ? An infinite plane, or the cube of which the given square constitutes a side. What is infinity for a cube ? Infinite three-dimensional space, or a figure of four dimensions.

Thus the usual concept of an infinite line remains, but to it there is added another, the concept of infinity as a plane resulting from the motion of the line in a direction perpendicular to itself.

The infinite three-dimensional sphere remains ; but a four-dimensional body constitutes infinity for a three-dimensional body.

Moreover, the problem becomes even simpler if we bear in mind that an " infinite " line, an " infinite " plane and an " infinite " solid are pure abstractions ; whereas a (finite) line in relation to a point, a square in relation to a line and a cube in relation to a square, are real concrete facts.

So, remaining within the domain of facts, the principle of infinity in geometry can be formulated as follows : for every figure of a given number of dimensions infinity is a figure of the given number of dimensions plus one.

At the same time the figure of the lower number of dimensions is *incommensurable* with the figure of the higher number of dimensions. Incommensurability (in figures of different numbers of dimensions) creates infinity.

All this is very elementary. But if we firmly bear in mind the inferences to be drawn from these elementary propositions, they will enable us to free ourselves from the influence of the wrongly interpreted Aristotelian principle of the constancy of phenomena. The principle of Aristotle is true only within the limits of the finite, within the limits of commensurability. As soon as the infinite begins, we know nothing and have no right to assert anything in relation to the unity of phenomena and laws.

Continuing these arguments, we meet with another still more interesting fact, that is, that *physical* infinity differs from *geometrical* infinity as greatly as geometrical infinity differs from *mathematical* infinity. Or, to be more precise, physical infinity begins *much sooner* than geometrical infinity. And if mathematical infinity has only one meaning and geometrical infinity two meanings, physical infinity can have many meanings, that is, the mathematical meaning (incomputability), the geometrical meaning (the presence of an additional dimension or immeasurable extension) and purely physical meanings, that is, difference in function.

Infinity is created by incommensurability. But incommensurability can be arrived at in different ways. And in the physical world incommensurability can be brought about because of the *quantitative* difference alone. As a rule, only quantities which are different qualitatively are regarded as incommensurable, and the qualitative difference is regarded as independent of the quantitative difference. But this is precisely where the chief mistake lies. Quantitative difference brings about qualitative difference.

In the mathematical world incommensurability is created by the *incomputability* of one of the quantities compared ; in the geometrical world it is created either by the infinite extension of one of the quantities which are being compared, or by the presence in it of a new dimension. In the physical world it is brought about simply by a difference in size which sometimes even permits of calculation.

All this means that infinity in geometry differs from infinity in mathematics in being *relative*. Mathematical infinity is equally infinity for any finite number. But geometrical infinity has no absolute meaning. A square is infinity for a line, but it is merely *bigger* than another smaller square or smaller than another bigger square.

In the physical world a large body is often incommensurable with a small one, and the small body bigger than the large one. A mountain is incommensurable with a mouse, and the mouse is *bigger* than the mountain by the perfection of its functions and by reason of its belonging to another level of being.

Further, it must be mentioned that *the function of every individual thing is possible only if the thing itself has a definite size.* The reason why this has not been noticed and established long ago is to be found in a wrong understanding of the principle of Aristotle.

Physicists have often come upon manifestations of this law, namely, that the function of every individual thing is possible only if the thing itself has a definite size, but it has never arrested their attention and never led them to put together observations obtained in different domains. In the formulation of many physical laws we find qualifications that the particular law is true only of medium quantities, and that in the case of larger quantities or smaller quantities the law changes. This law is still more clearly seen in the phenomena studied by biology and sociology.

The conclusion from what has been said can be formulated in the following way :

All that exists is what it is only within the limits of a certain and very restricted scale. On a different scale it becomes something else. In other words, every thing and every event has a certain meaning only within the limits of a certain scale, when compared with things and events of proportions not very far removed from its own, that is, existing within the same scale.

A chair cannot be a chair in the planetary world. Similarly, a chair cannot be a chair in the world of electrons. A chair has its meaning and its three dimensions only among objects created by the hand of man, serving the needs and requirements of man, and commensurable with man. On the planetary scale a chair cannot have individual existence because it cannot have any function. It is simply a small particle of matter inseparable from the matter surrounding it. As has been explained before, in the world of electrons *also* a chair becomes too small for its function and therefore loses all its meaning and all its significance. A *chair* actually does not exist in comparison even with things which differ from it much less than planets or electrons. A chair in the midst of the ocean, or a chair in the midst of the Alpine ranges, would be a point having no dimension.

All this shows that incommensurability exists not only among things of different categories and denominations, and not only among things of a different number of dimensions, but also among things which merely differ considerably in size. A big object is incommensurable with a small object. A big object is often infinite in comparison with a small one.

Every separate thing and every separate phenomenon, in becoming bigger or smaller, ceases to be what it was and becomes something else—something belonging to another category.

This principle is still utterly foreign to physics, both to the old and to the new. On the contrary, every separate thing and every separate phenomenon remains for physics what it was originally recognised as being—matter remains matter, motion remains motion, velocity remains velocity. And yet it is precisely this possibility of the transition of space phenomena into time phenomena and of time phenomena into space phenomena which conditions the eternal fluctuation of life. And this transition takes place when the given phenomenon becomes infinity in relation to another phenomenon.

From the point of view of old physics, velocity, which was considered a generally understood phenomenon requiring no definition, always remained velocity; it could grow, increase, become an *infinite velocity*. It occurred to no one to doubt it. And having only accidentally stumbled upon the fact that the velocity of light is a limiting velocity, physicists were forced to admit that all was not well, and that the idea of velocity needed revision.

But physicists certainly could not surrender at once and admit that velocity can cease to be velocity and can become something else.

What did they actually stumble upon?

They stumbled upon an instance of infinity. The velocity of light is infinity as compared with all the velocities which can be observed or created experimentally. And, as such, it cannot be increased. In actual fact it ceases to be velocity and becomes an *extension*.

A ray of light possesses an additional dimension as compared with any object moving with " terrestrial velocities ".

A line is infinity in relation to a point. And the motion of the point does not alter this relation; a line will always remain a line.

The idea of limiting velocity presented itself when physicists hit upon a case of obvious infinity. But even apart from this, all the inconsistencies and contradictions in the old physics which were discovered and calculated by Prof. Einstein and supplied him with material for the building of his theories, all these without exception result from the difference between the infinite and the finite. He himself often alludes to this.

Einstein's description of the example of " the behaviour of clocks and measuring rods on a rotating marble disc " suffers from one defect. Prof. Einstein forgot to say that the diameter of the " marble disc " to which are fastened the clocks which begin to go at different speeds with the movement of the disc, according to their distance from the centre of the disc, should be approximately equal to the distance from the earth to Sirius ; or else, the " clocks " must be the size of an atom (about five million of which can be put in the diameter of a full-stop). With such a difference in size strange phenomena can actually occur, such as the unequal speed of the clocks or the change in the length of the measuring rods. But there could not be a " disc " with a diameter equal to the distance from the earth to Sirius, or clocks the size of an atom. Such clocks will cease to exist before they change their speed, though this cannot be intelligible to modern physics, which, as I pointed out before, cannot get free of the Aristotelian principle of the constancy of phenomena and cannot therefore notice that constancy is always destroyed by incommensurability. It can be assumed generally that within the limits of terrestrial possibilities the behaviour of both the clocks and the measuring-rods will be quite respectable, and for all practical purposes we can safely rely upon them. There is only one thing we must not do—we must not set them any " problems with infinity ".

After all, all the misunderstandings are caused by problems with infinity, chiefly because infinity is introduced on a level with finite quantities. The result will of course be different from what is expected ; an unexpected result demands adaptation. The " special principle of relativity " and the " general principle of relativity " are very complicated and cumbersome adaptations for the explanation of the strange and unexpected results of " problems with infinity ".

Prof. Einstein himself writes that proofs of his theories can be found either in astronomical phenomena or in the phenomena of electricity and light. In other words, he affirms by this that all problems that require particular principles of relativity arise from problems with infinity or with incommensurability.

The special principle of relativity is based on the difficulty of establishing the simultaneity of two events separated by space, and above all on the impossibility of the composition of velocities in comparing terrestrial velocities with the velocity of light. This is precisely a case of the established heterogeneity of the finite and the infinite.

Of this heterogeneity I have spoken earlier; as regards the impossibility of establishing the simultaneity of two events Prof. Einstein does not specify at what distance between two events the establishment of their simultaneity becomes impossible. And if we insist upon an explanation we shall certainly receive the answer that the distance must be " very great ". This " very great " distance again shows that Prof. Einstein presumes a problem with infinity.

Time is really different for different moving systems of bodies. But it is incommensurable (or it cannot be synchronised) only if the moving systems are separated by a large space which is actually infinity for them, or when they differ greatly in size or velocity, that is, when one of them is infinity in comparison with the other, or contains infinity.

And to this may be added that not only time, but also space, is different for them, changing according to their size and velocity.

The general proposition is quite correct—

" Every separately existing system has its own time."

But what does " separately existing " mean ? And how can there be separate systems in a *world of connected spirals* ? All that exists in the world constitutes one whole ; there can be nothing separate.

The principle of the absence of separateness, of the impossibility of separateness, constitutes a very important part of certain philosophical teachings, for instance of Buddhism, where one of the first conditions for a right understanding of the world is considered to be the destruction of the " sense of separateness " in oneself.

From the point of view of the new model of the universe separateness exists, but only relatively.

Let us imagine a system of cog-wheels, rotating with different velocities, which depend upon their size and upon the place occupied by each of them in the system. The system, for instance the mechanism of an ordinary watch, constitutes one whole, and from this point of view there can be nothing separate in it. From another point of view each separate cog-wheel moves at its own velocity, i.e. it has a separate existence and *its own time*.

In analysing the problem of *infinity* and infinite quantities we touch upon several other problems, the elucidation of which is equally necessary for a right understanding of the new model of the universe. Some of these problems have already been examined. There remain the problems of *zero quantities* and *negative quantities*.

Let us try to begin the examination of these quantities in the same way as we began the examination of infinity and infinite quantities, that is, let us try to compare their meanings in mathematics, in geometry and in physics.

Zero in mathematics has always one meaning. There is no reason to speak of *zero quantities* in mathematics.

Zero in mathematics and the point in geometry have approximately the same meaning, with the difference that the point in geometry indicates the *place* at which something begins or at which something ends, or at which something happens, for instance where two lines intersect one another; whereas in mathematics zero indicates the limit of certain possible operations. But in their essence there is no difference between zero and the point, because neither has independent existence.

The case is quite different in physics. The *material point* is a point only on a given scale. If the scale is changed the point can prove to be a very complex and many-dimensional system of immense measurements.

Let us imagine a small map on which even the biggest towns are points. Let us suppose that we have found the means to bring out the content of these points or to fill them with content. Then, what looked like a point will manifest a great many new properties and characteristics, and the extensions and measurements included in it. In the town will appear streets, parks, houses, people. How are the measurements of these streets, squares and people to be understood ?

When the town was for us a point, they were *smaller than a point*. Is it not possible to call them *negative dimensions* ?

The uninitiated, in most cases, do not know that the concept "*negative quantity*" has no definition in mathematics. It has a certain meaning only in elementary arithmetic, and also in algebraical formulæ, where it designates the *operation* to be performed, rather than the difference in the properties of the quantities. In physics "negative quantity" does not mean anything at all. Nevertheless we have already come upon negative quantities. It was when speaking of dimensions inside the atom, that I had to point out that although the atom (or the molecule) has no dimension for the direct senses, i.e. is equal to zero, these dimensions or extensions *inside the atom* are still smaller, i.e. *smaller than zero*.

So we need no metaphors or analogies in order to speak about negative dimensions. These are the dimensions within what appears to be a material point. And this explains exactly why it is wrong to regard small particles of matter such as atoms or electrons as material. They are not material, because they are *negative* physically, i.e. smaller than *physical* zero.

Putting together all that has been set forth hitherto, we see that besides the *period of six dimensions*, we have *imaginary dimensions*, the seventh, the eighth and so on, which proceed in non-existent directions and differ in the degree of impossibility ; and *negative dimensions* within the smallest particles representing for us material points.

In new physics the conflict between the old and the new ideas of time and space is especially marked in conceptions as to the ray of light, but at the same time a right understanding of the ray of light will solve all points at issue in the question of time and space.

I will complete the new model of the universe by an analysis of a ray of light, but before beginning this analysis I must examine certain further properties of time taken as a three-dimensional continuum.

Until now I have taken time as the measure of motion. But motion in itself is the sensation of an *incomplete* perception of the space in question. For a dog, for a horse, for a cat, our third dimension is motion. For us motion begins in the fourth dimension and is a partial sensation of the fourth dimension. But as for animals the imaginary movements of objects which in reality constitute their third dimension merge into those movements which are movements for us, that is into the fourth dimension, so for us movements of the fourth dimension merge into movements of the fifth and sixth dimensions. Starting from this we must endeavour to establish something which will allow us to judge the properties of the fifth and sixth dimensions. Their relation to the fourth dimension must be analogous to the relation of the fourth dimension to the third, of the third to the second, and so on. This means that first of all the new, the higher, dimension must be incommensurable with the lower dimension and form infinity for it, seeming to repeat its characteristics an infinite number of times.

Thus, if we take " time " (that is extension from before to after) as the fourth dimension, what will be the fifth dimension in this case, that is, what forms infinity for time, what is incommensurable with time ?

It is precisely phenomena of light that enable us to come into immediate contact with movements of the fifth and sixth dimensions.

The line of the fourth dimension is always and everywhere a closed curve, although on the scale of our three-dimensional perception we do not see either that this line is curved or that it is closed. This closed curve of the fourth dimension, or the circle of time, is the life or existence of every separate object, of every separate system, which is examined in time. But the circle of time does not break up or disappear. It continues to exist, and joining other, previously formed circles, it passes into eternity. Eternity is the infinite repetition of the completed circle of life, an infinite repetition of *existence*. Eternity is incommensurable with time. Eternity is infinity for time.

Quanta of light are precisely such circles of *eternity*.

The third dimension of time (the sixth dimension of space) is the stretching out of these eternal circles into a spiral or a cylinder with a screw-thread in which each circle is locked in itself (and motion along it is eternal) and simultaneously passes into another circle which is also eternal, and so on.

This hollow cylinder with two kinds of thread would be a model of a ray of light—a model of three-dimensional time.

The next question is, where is the electron? What happens to the electron of the luminous molecule which sends out quanta of light? This is one of the most difficult questions for new physics.

From the point of view of the new model of the universe the answer is clear and simple.

The electron is transformed into quanta, it becomes a ray of light. The point is transformed into a line, into a spiral, into a hollow cylinder.

As three-dimensional bodies electrons do not exist for us. The fourth dimension of electrons, that is *their existence* (the completed circle), also has no measurement for us. It is too small, has too short duration, is shorter than our thought. We cannot know about them, i.e. we cannot perceive them in a direct way.

Only the fifth and sixth dimensions of electrons have certain measurements in our space-time. The fifth dimension constitutes the thickness of the ray, and the sixth dimension its length.

Therefore in radiant energy we deal not with electrons themselves, but with their time dimensions, with the traces of their movement and existence, of which the primary web of any matter is woven.

Now if we accept the approximate description of the ray of light as a hollow cylinder consisting of quanta lying close to one another lengthwise along the ray, the picture becomes clearer.

First of all, the conflict between the theory of undulatory movements and the emission theory is settled, and it is settled in the sense that both theories prove to be equally true and equally necessary, though they refer to different phenomena or to different sides of the same kind of phenomenon.

Vibrations or undulatory movements, which were taken for the cause of light, are undulatory movements transmitted *along already existing rays of light*. What is called the " velocity of light " is probably the velocity of these vibrations passing along the ray. This explains why the calculations made on the basis of the theory of vibrations proved to be correct and made new discoveries possible. In itself a ray has no velocity; it is a line, a space concept, not a time concept.

No æther is necessary, for vibrations travel *by light itself*. At the same time light has " atomic structure ", for a cross-section of a beam of light would show a network through the mesh of which the molecules of the gas it meets can easily slip.

In spite of the fact that scientists speak of the very accurate methods which they possess for counting electrons and measuring their velocities, it is permissible to have doubts whether they really mean electrons and not their extensions along the sixth dimension, the extensions which have already acquired space meaning for us.

The material structure of a ray of light explains also its possible deviations under the influence of forces acting upon it. But it is certain that these forces are not " attraction " in the Newtonian sense, although they may very possibly be magnetic attraction.

There still remains one question I have intentionally left untouched until now. This is the question of the duration of the existence of small particles, molecules, atoms and electrons. This question has never received serious consideration in physics; small units are regarded as *constant*, like matter and energy, that is, as existing for an indefinitely long time. If there were ever any doubts about this, they have not left a noticeable trace, and physicists speak of molecules, atoms and electrons, first (as has already been pointed out), as *particles of matter*, and, second, as particles which exist parallel with ourselves, occupying a certain time within our time. This is never said directly, but on this point doubt never arises. And yet in reality the existence of small units of matter is so short that it is quite impossible to speak of them in the same language as that in which we speak of physical bodies when they are the subject of our examination.

It was made clear before that the space of small units is proportionate to their size, and in exactly the same way their time is proportionate to their size. This means that their time, i.e. the time of their existence, is almost non-existent in comparison with our time.

Physics speaks of observing electrons and calculating their weight, velocity, etc. But an electron is for us only a *phenomenon*, and a phenomenon which is quicker than anything visible to our eyes; an atom as a whole is perhaps only a longer phenomenon, but longer on the same scale, just as there are various instantaneous speeds in a photographic camera. But both the atom and the electron are only time phenomena for us and, moreover, " instantaneous " phenomena;

they are not bodies, not objects. Some scientists assert that they have succeeded in seeing molecules. But do they know how long by their clock a molecule can exist ? During its very short existence, a molecule of gas (which alone may be accessible to observation, if this be possible at all) travels through immense distances and will in no case appear either to our eye or to the photographic camera as a moving point. And seen as a line it would inevitably intersect with other lines, so that it would be more than difficult to trace a single molecule, even for the period of a small fraction of a second ; and even if this became possible in some way it would require such magnification as is actually impossible up to the present time.

All this must be kept in view in speaking, for instance, of phenomena of light. A great many misunderstandings fall away at once if we realise and carefully bear in mind the fact that an " electron " exists for an immeasurably small part of a second, which means that it can never under any condition be seen or measured by us, as we are.

It is impossible with existing *scientific* material to find firm ground for any theory as to the short existence of small units of matter. The material for such a theory is to be found in the idea of " different time in different cosmoses ", which forms part of a special doctrine on the world, which will be the subject of another book.

1911–1929.

CHAPTER II
THE FOURTH DIMENSION

The idea of hidden knowledge—The problem of the invisible world and the problem of death—The invisible world in religion, in philosophy, in science—The problem of death and various interpretations of it—The idea of the fourth dimension—Various approaches to it—Our position in relation to the "domain of the fourth dimension" —Methods of studying the fourth dimension—Hinton's ideas—Geometry and the fourth dimension—Morosoff's article—An imaginary world of two dimensions—The world of perpetual miracle—The phenomena of life—Science and unmeasurable phenomena— Life and thought—Perception of plane-beings—A plane-being's different stages of under- standing the world—Hypothesis of the third dimension—Our relation to the "invisible" —The world of the unmeasurable round us—Unreality of bodies of three dimensions —Our own fourth dimension—Deficiency of our perception—Properties of perception in the fourth dimension—Inexplicable phenomena of our world—The psychic world and attempts to interpret it—Thought and the fourth dimension—Expansion and contrac- tion of bodies—Growth—The phenomena of symmetry—Diagrams of the fourth dimension in Nature—Movement from the centre along radii—The laws of symmetry—States of matter—Relation of time and space in matter—Theory of dynamic agents—Dynamic character of the universe—The fourth dimension within us—The "Astral sphere"— Hypothesis of fine states of matter—Transformation of metals—Alchemy—Magic— Materialisation and dematerialisation—Prevalence of theories and absence of facts in astral hypotheses—Necessity for a new understanding of "space" and "time".

THE idea of the existence of a hidden knowledge, surpassing all the knowledge a man can attain by his own efforts, must grow and strengthen in people's minds from the realisation of the insolu- bility of many questions and problems which confront them.

Man may deceive himself, may think that his knowledge grows and increases, that he knows and understands more than he knew and understood before, but sometimes he may be sincere with himself and see that in relation to the fundamental problems of existence he is as helpless as a savage or a little child, although he has invented many clever machines and instruments which have complicated his life but have not rendered it any more comprehensible.

Speaking still more sincerely with himself man may recognise that all his scientific and philosophical systems and theories are similar to these machines and implements, for they only serve to complicate the problems without explaining anything.

Among the insoluble problems with which man is surrounded, two occupy a special position—the problem of the invisible world and the problem of death.

In all the history of human thought, in all the forms, without exception, which this thought has ever taken, people have always

95

divided the world into the *visible* and the *invisible*; and they have always understood that the visible world accessible to their direct observation and study represents something very small, perhaps even something non-existent, in comparison with the enormous existent invisible world.

Such an assertion, that is, that the division of the world into the visible and the invisible has existed always and everywhere, may appear strange at first, but in reality all existing general schemes of the world, from the most primitive to the most subtle and elaborate, divide the world into the visible and the invisible and can never free themselves from this division. This division of the world into the visible and the invisible is the foundation of man's thinking about the world, no matter how he names or defines this division.

The fact of such a division becomes evident if we try to enumerate the various systems of thinking about the world.

First of all let us divide all the systems of thinking about the world into three categories :

1. Religious systems.
2. Philosophical systems.
3. Scientific systems.

All religious systems without exception, from those theologically elaborated down to the smallest details, such as Christianity, Buddhism, Judaism, to the completely degenerated religions of " savages ", appearing as " primitive " to modern knowledge, invariably divide the world into visible and invisible. In Christianity : God, angels, devils, demons, souls of living and dead people, heaven or hell. In paganism : gods personifying forces of nature, thunder, sun, fire, spirits of mountains, woods, lakes, water-spirits, house spirits—all this is the invisible world.

In philosophy there is the world of events and the world of causes, the world of things and the world of ideas, the world of phenomena and the world of noumena. In Indian philosophy, especially in certain schools of it, the visible or phenomenal world, that is, Maya or illusion, which means a wrong conception of the invisible world, does not exist at all.

In science, the invisible world is the world of small quantities and, strange though it is, also the world of large quantities. The visibility of the world is determined by the scale. The invisible world is on the one hand the world of micro-organisms, cells, the microscopic and the ultra-microscopic world ; still further it is the world of molecules, atoms, electrons, " vibrations ", and, on the other hand, the world of invisible stars, other solar systems, unknown universes. The microscope expands the limits of our vision in one direction, the telescope in the other. But both increase visibility very little in comparison with what remains invisible. Physics and chemistry show us the possibility of investigating phenomena in such small quantities or in such distant worlds as will never be visible to us. But this only strengthens the fundamental idea of the existence of an enormous, invisible world round the small, visible world.

Mathematics goes even farther As was pointed out before, it calculates such relations of magnitudes and such relations between these relations as have nothing similar in the visible world surrounding us. And we are forced to admit that the *invisible* world differs from the visible not only in size, but in some other properties which we can neither define nor understand and which only show us that laws, inferred by us for the visible world, cannot refer to the invisible world.

In this way invisible worlds, the religious, the philosophical, and the scientific, are, after all, more closely related to one another than they would at first appear. And these invisible worlds of different categories possess identical properties common to all. These properties are, first : incomprehensibility for us, that is, incomprehensibility from the ordinary point of view, or for ordinary means of cognition ; and, second : the fact that they contain the causes of the phenomena of the visible world.

This idea of causes is always associated with the invisible world. In the invisible world of the religious systems, invisible forces govern people and visible phenomena. In the scientific invisible world the causes of visible phenomena always come from the invisible world of small quantities and " vibrations ". In philosophical systems the phenomenon is only our conception of the noumenon, that is, an illusion, the real cause of which remains hidden and inaccessible to us.

This shows that on all levels of his development man has always understood that the causes of visible and observable phenomena lie beyond the sphere of his observation. He has found that among observable phenomena certain facts could be regarded as causes of other facts, but these deductions were insufficient for the explanation of *everything* that occurred in himself and around him. Therefore in order to be able to explain the causes it was necessary for him to have an invisible world consisting either of " spirits ", or " ideas ", or " vibrations ".

The other problem which attracted the attention of men by its insolubility and which by the form of its approximate solution determined the direction and development of human thought, was the problem of death, that is, the explanation of death, the idea of future

life, of the immortal soul, or the absence of the immortal soul, and
so on.

Man could never reconcile himself to the idea of death as disappearance. Too many things contradicted it. There were in
himself too many traces of the dead, their faces, words, gestures,
opinions, promises, threats, the feelings which they roused, fear,
jealousy, desire. All these continued to live in him, and the fact of
their death was more and more forgotten. A man saw his dead
friend or enemy in his dreams. He appeared exactly as he was before.
Evidently he was living *somewhere*, and could come *from somewhere*
by night.

So it was very difficult to believe in death and man always needed
theories for the explanation of existence after death.

On the other hand, echoes of esoteric teachings on life and death
sometimes reached man. He could hear that the visible, earthly,
observable, life of man is only a small part of the life belonging to
him. And man of course understood in his own way these fragments which reached him, changed them in his own fashion, adapted
them to his own level and understanding, and built out of them some
theory of future existence, similar to existence on the earth.

The greater part of religious teachings on the future life connect it
with the idea of reward or punishment, sometimes in an undisguised,
sometimes in a veiled, form. Heaven and hell, transmigration of
souls, reincarnation, the wheel of lives—all these theories contain the
idea of reward or punishment.

But religious theories often do not satisfy man, and in addition
to the recognised, orthodox, ideas of life after death there usually
exist other, as it were illegitimate, ideas of the world beyond the grave
or of the spirit-world, which allow a greater freedom of imagination.

No religious teaching, no religious system, can by itself satisfy
people. There is always some other, more ancient system of popular
belief underlying it or hiding behind it. Behind external Christianity,
behind external Buddhism, there stand the remains of ancient pagan
creeds (in Christianity the remains of pagan beliefs and customs, in
Buddhism " the cult of the devil "), which sometimes make a deep
mark on the external religion. In modern Protestant countries, for
instance, where the remains of ancient paganism are already completely
extinct, there have come into existence, under the outward mask of
logical and moral Christianity, systems of primitive thinking of the
world beyond the grave, such as spiritualism and kindred teachings.

And theories of existence beyond the grave are always connected
with theories of the invisible world ; the former are always based
upon the latter.

This all relates to religion and "pseudo-religion". There are no philosophical theories of existence beyond the grave. All theories of life after death can be called religious or, more correctly, pseudo-religious.

Moreover, it is difficult to take philosophy as a whole, so diverse and contradictory are the various speculative systems. Still, to a certain extent, it is possible to accept as a standard of philosophical thinking the view which can see the unreality of the phenomenal world and the unreality of man's existence in the world of things and events, the unreality of the separate existence of man and the incomprehensibility for us of the forms of real existence, although this view can be based on very different foundations, either materialistic or idealistic. In both cases the question of life and death acquires a new character and cannot be reduced to the naïve categories of ordinary thinking. For such a view there is no particular difference between life and death, because, strictly speaking, for it there are no proofs of a separate existence, of separate lives.

There are not and there cannot be any *scientific* theories of existence after death because there are no facts in favour of the reality of such an existence, while science, successfully or unsuccessfully, wishes to deal with facts. In the fact of death the most important point for science is a certain change in the state of the organism, which stops all vital functions, and the decomposition of the body following upon it. Science sees in man no psychic life independent of the vital functions, and all theories of life after death, from the scientific point of view, are pure fiction.

Modern attempts at "scientific" investigation of spiritualistic phenomena and similar things lead nowhere and can lead nowhere, for there is a mistake here in the very setting of the problem.

In spite of the difference between the various theories of the future life, they all have one common feature. They either picture the life beyond the grave as similar to the earthly life, or deny it altogether. They do not and cannot attempt to conceive life after death in new forms or new categories. And this is precisely what makes all usual theories of life after death unsatisfactory. Philosophical and strictly scientific thought shows us the necessity of reconsidering the problem from completely new points of view. A few hints coming from the esoteric teaching partly known to us indicate the same.

It already becomes evident that if the problem of death and life after death can be approached in any way, it must be approached from quite a new angle. In the same way, the question of the invisible world must also be approached from a new angle. All we know, all we have thought till now, shows us the reality and the vital importance of these problems. Until he has answered in one way or another the questions of the invisible world and of life after death, man cannot think of anything else without creating a whole series of contradictions. Right or wrong, man must construct for himself some kind of explanation. And he must base his treatment of the problem of death either upon science, or upon religion, or upon philosophy.

But to a thinking man both the " scientific " denial of the possibility of life after death and the pseudo-religious admission of it, (for we know nothing but pseudo-religion), as well as different spiritualistic, theosophical and similar theories, quite justly appear equally naïve.

Nor can the abstract philosophical view satisfy man. Such a view is too remote from life, too remote from direct, real sensations. One cannot live by it. In relation to the phenomena of life and their possible causes, unknown to us, philosophy is very like astronomy in its relation to the distant stars. Astronomy calculates the movement of stars which are at colossal distances from us. But all celestial bodies are alike for it. They are nothing but moving dots.

Thus, philosophy is too remote from concrete problems such as the problem of future life. Science does not know the world beyond the grave ; pseudo-religion creates the other world in the image of the earthly world.

This helplessness of man in the face of the problems of the invisible world and of death becomes particularly obvious when we begin to realise that the world is far bigger and far more complex than we have hitherto thought, and that what we think we know occupies only a very insignificant place amidst that which we do not know.

Our basic conception of the world must be broadened. Already we feel and know that we can no longer trust the eyes with which we see, or the hands with which we touch. The real world eludes us at such attempts to ascertain its existence. A more subtle method, a more efficient means, are needed.

The ideas of the " fourth dimension ", ideas of " many dimensional space ", show the way by which we may arrive at the broadening of our conception of the world.

The expression " fourth dimension " is often met with in conversational language and in literature, but it is very seldom that anybody has a clear idea of what it really means. Generally the fourth dimension is used as the synonym of the mysterious, miraculous, " supernatural ", incomprehensible and incognisable, as a kind of general definition of the phenomena of the " super-physical " world.

" Spiritualists " and " occultists " of various schools often make use of this expression in their literature, assigning to the sphere of the fourth dimension all the phenomena of the " world beyond " or the " astral sphere ". But they do not explain what it means, and from what they say one can understand only that the chief property which they ascribe to the fourth dimension is " unknowableness ".

The connecting of the idea of the fourth dimension with existing theories of the invisible world or the world beyond is certainly quite fantastic, for, as has already been said, all religious, spiritualistic, theosophical and other theories of the invisible world first of all make it exactly similar to the visible and consequently " three-dimensional " world.

Therefore mathematics quite justly objects to the established view of the fourth dimension as something belonging to the " beyond ".

The very idea of the fourth dimension must have arisen in close connection with mathematics, or, to put it better, in close connection with the idea of measuring the world. It must have arisen from the supposition that, besides the three known dimensions of space—length, breadth and height—there might also exist a fourth dimension, inaccessible to our perception.

Logically, the supposition of the existence of the fourth dimension can be based on the observation of those things and events in the world surrounding us for which the measurement in length, breadth and height is not sufficient, or which elude all measurement; because there are things and events the existence of which calls for no doubt, but which cannot be expressed in any terms of measurement. Such are, for instance, various effects of vital and psychic processes ; such are all ideas, mental images and memories ; such are dreams. If we consider them as existing in a real, objective sense, we can suppose that they have some other dimension besides those accessible for us, that is, some extension immeasurable for us.

There exist attempts at a purely mathematical definition of the fourth dimension. It is said for instance : " In many problems of pure and applied mathematics formulæ and mathematical expressions are met with containing four or more variable quantities, each of which, independently of the others, may be positive or negative and lie between $+ \infty$ and $- \infty$. And as every mathematical formula, every equation, can have a dimensional expression, so from this is deduced an idea of space which has four or more dimensions ".[1]

The weak point of this definition is the proposition accepted as unquestionable that every mathematical formula, every equation, can have a dimensional expression. In reality such a proposition is entirely without ground, and this deprives the definition of all meaning.

---

[1] The article " Four-dimensional space " in the Russian Encyclopedia of Brockhaus and Efron.

Reasoning by analogy with the existing dimensions, it must be supposed that if the fourth dimension existed it would mean that side by side with us lies some other space which we do not know, do not see, and into which we are unable to pass. It would then be possible to draw a line from any point of our space into this " domain of the fourth dimension " in a direction unknown to us and impossible either to define or to comprehend. If we could visualise the direction of this line going out of our space then we should see the " domain of the fourth dimension ".

Geometrically this proposition has the following meaning. We can conceive simultaneously three lines perpendicular and not parallel to one another. These three lines are used by us to measure the whole of our space, which is therefore called three-dimensional. If the " domain of the fourth dimension " lying outside our space exists, this means that besides the three perpendiculars known to us, determining the length, the breadth and the height of solids, there must also exist a fourth perpendicular, determining some new extension unknowable to us. Then the space measurable by these four perpendiculars could be called four-dimensional.

We are unable to define geometrically, or to conceive, this fourth perpendicular, and the fourth dimension still remains extremely enigmatic. The opinion is sometimes met with that mathematicians know something about the fourth dimension which is inaccessible to ordinary mortals. Sometimes it is said, and one can even find such assertions in literature, that Lobatchevsky " discovered " the fourth dimension. During the last twenty years the discovery of the " fourth dimension " has often been ascribed to Einstein or Minkovsky.

In reality mathematics can say very little about the fourth dimension. There is nothing in the hypothesis of the fourth dimension that would make it inadmissible from a mathematical point of view. This hypothesis does not contradict any of the accepted axioms and, because of this, does not meet with particular opposition on the part of mathematics. Mathematicians even admit the possibility of establishing the relationship that should exist between four-dimensional and three-dimensional space, i.e. certain properties of the fourth dimension. But they do all this in a very general and rather indefinite form. No exact definition of the fourth dimension exists in mathematics.

Lobatchevsky actually treated the geometry of Euclid, i.e. geometry of three-dimensional space, as a particular case of geometry, which ought to be applicable to a space of any number of dimensions. But this is not mathematics in the strict sense of the word, it is only metaphysics on mathematical themes; and the deductions from it cannot be formulated mathematically or can be formulated only in specially constructed conditional expressions.

Other mathematicians regarded axioms accepted in the geometry of Euclid as artificial and incorrect, and attempted to disprove them on the strength, chiefly, of certain deductions from Lobatchevsky's spherical geometry, and to prove, for instance, that parallel lines meet. They contended that the accepted axioms are correct only for three-dimensional space, and on the basis of their arguments, which disproved these axioms, they built up a new geometry of many dimensions.

But all this is not geometry of four dimensions.

The fourth dimension could only be considered as geometrically proved when the direction of the unknown line starting from any point of our space and going into the region of the fourth dimension could be determined, i.e. when a means of constructing a fourth perpendicular is found.

It is difficult to describe even approximately the significance which the discovery of the fourth perpendicular in our universe would have for our knowledge. The conquest of the air ; hearing and seeing at a distance ; establishing connections with other planets or with other solar systems ; all this is nothing in comparison with the discovery of a new dimension. But so far it has not been made. We must recognise that we are helpless before the riddle of the fourth dimension, and we must try to examine the problem within the limits accessible to us.

After a closer and more exact investigation of the problem itself we come to the conclusion that it cannot be solved in existing conditions. The problem of the fourth dimension, though purely geometrical at the first glance, cannot be solved by geometrical means. Our geometry of three dimensions is as insufficient for the investigation of the question of the fourth dimension as planimetry alone is insufficient for the investigation of questions of stereometry. We must find the fourth dimension, if it exists, in a purely experimental way, and also find a means for a projective representation of it in three-dimensional space. Only then shall we be able to create a geometry of four dimensions.

Even slight acquaintance with the problem of the fourth dimension shows the necessity for studying it from the psychological and physical sides.

The fourth dimension is unknowable. If it exists and if at the same time we cannot know it, it evidently means that something is lacking in our psychic apparatus, in our faculties of perception ; in other words, phenomena of the region of the fourth dimension are not reflected in our organs of sense. We must examine why this should be so, what are our defects on which this non-receptivity depends, and must find the conditions (even if only theoretically) which would make the fourth dimension comprehensible and accessible to us. These are all questions relating to psychology or, possibly, to the theory of knowledge.

Further, we know that the region of the fourth dimension (again, if it exists) is not only unknowable for our psychic apparatus, but is *inaccessible* in a purely physical sense. This must depend not on our defects, but on the particular properties and conditions of the region of the fourth dimension itself. It is necessary to examine what these conditions are, which make the region of the fourth dimension inaccessible to us, and to find the relation between the physical conditions of the region of the fourth dimension and the physical condition of our world. And having established this, it is necessary to see whether in the world surrounding us there is anything similar to these conditions, that is, whether there are any relations analogous to relations between the region of three dimensions and that of four dimensions.

Speaking in general, before attempting to build up a geometry of four dimensions it is necessary to create a physics of four dimensions, that is, to find and to define physical laws and conditions which may exist in the space of four dimensions.

Many people have worked at the problem of the fourth dimension.

Fechner wrote a great deal about the fourth dimension. From his discussions about worlds of one, two, three and four dimensions there follows a very interesting method of investigating the fourth dimension by means of building up analogies between worlds of different dimensions, i.e. between an imaginary world on a plane and the three-dimensional world, and between the three-dimensional world and the world of four dimensions. This method is used by nearly all those who have ever studied the problem of higher dimensions, and we shall have occasion to meet with it further on.

Professor Zöllner evolved the theory of the fourth dimension from observations of " mediumistic " phenomena, chiefly of phenomena of so-called " materialisation ". But his observations have long been considered doubtful because of the established fact of the insufficiently strict arrangement of his experiments (Podmore and Hislop).

A very interesting summary of almost all that has ever been written about the fourth dimension up to the nineties of last century is to be found in the books of C. H. Hinton. These books contain also many of Hinton's own ideas ; but, unfortunately, side by side with the valuable ideas there is a great deal of unnecessary dialectic such as always accumulates round the question of the fourth dimension.

Hinton makes several attempts at a definition of the fourth dimension from the physical side, as well as from the psychological. Considerable space is occupied in his books by the description of a method, invented by him, of accustoming the mind to cognition of the fourth dimension. It consists of a long series of exercises for the perceiving and the visualising apparatus, with sets of differently coloured cubes, which are meant to be memorised, first in one position, then in another, then in a third, and after that to be visualised in different combinations.

The fundamental idea which guided Hinton in the creation of this method of exercises is that the awakening of " higher consciousness " requires the " casting out of the self " in the visualisation and cognition of the world, i.e. the accustoming of oneself to know and conceive the world, not from a personal point of view (as we generally know and conceive it), but as it is. For this it is necessary, first of all, to learn to visualise things not as they appear to us, but as they are, even if only in a geometrical sense ; from this there must develop the capacity to know them, i.e. to see them, as they are, also from other points of view besides the geometrical.

The first exercise suggested by Hinton consists in the study of a cube composed of 27 smaller cubes coloured differently and bearing definite names. After having thoroughly learned the cube made up of smaller cubes, it has to be turned over and learned and memorised in the reverse order. Then the position of the smaller cubes has to be changed and memorised in that order, and so on. As a result, according to Hinton, it is possible to cast out in the cube studied the concepts " up and down ", " right and left ", and so on, and to know it independently of the position with regard to one another of the smaller cubes composing it, i.e. probably to visualise it simultaneously in different combinations. This would be the first step towards casting out the self-elements in the conception of the cube. Further on, there is described an elaborate system of exercises with series of differently coloured and differently named cubes, out of which various figures are composed. All this has the same purpose, to cast out the self-elements in the percepts and in this way to develop higher consciousness.

Casting out the self-elements in percepts, according to Hinton's idea, is the first step towards the development of higher consciousness and towards the cognition of the fourth dimension.

He says that if there exists the capacity of vision in the fourth dimension, that is, if we are able to see objects of our world as if from the fourth dimension, then we shall see them, not as we see them in the ordinary way, but quite differently.

We usually see objects as either above or below us, or on the same level with us, to the right or to the left, behind us or in front of us, and always from one side only—the one facing us—and in perspective. Our eye is an extremely imperfect instrument; it gives us an utterly incorrect picture of the world. What we call perspective is in reality a distortion of visible objects which is produced by a badly constructed optical instrument—the eye. We see all objects distorted. And we visualise them in the same way. But we visualise them in this way entirely owing to the habit of seeing them distorted, that is, owing to the habit created by our defective vision, which has weakened the capacity of visualisation.

But, according to Hinton, there is no necessity to visualise objects of the external world in a distorted form. The power of visualisation is not limited by the power of vision. We see objects distorted, but we know them as they are. And we can free ourselves from the habit of visualising objects as we see them, and we can learn to visualise them as we know they really are. Hinton's idea is precisely that before thinking of developing the capacity of seeing in the fourth dimension, we must learn to visualise objects as they would be seen from the fourth dimension, i.e. first of all, not in perspective, but from all sides at once, as they are known to our "consciousness". It is just this power that should be developed by Hinton's exercises. The development of this power to visualise objects from all sides at once will be the casting out of the self-elements in mental images. According to Hinton, "casting out the self-elements in mental images must lead to casting out the self-elements in perceptions". In this way, the development of the power of visualising objects from all sides will be the first step towards the development of the power of seeing objects as they are in a geometrical sense, i.e. the development of what Hinton calls a "higher consciousness".

In all this there is a great deal that is right, but also a great deal that is arbitrary and artificial. First of all, Hinton does not take into consideration the difference between the various psychic types of men. A method that may prove satisfactory for himself may produce no results or even contrary results for other people. Second, the very psychological foundation of his system of exercises is too unstable. Usually he does not know when to stop, carries his analogies too far and in that way deprives many of his conclusions of all value.

From the point of view of geometry, according to Hinton, the question of the fourth dimension may be examined in the following way.

We know geometrical figures of three kinds :

Figures of one dimension—lines.

Figures of two dimensions—planes.

Figures of three dimensions—solids.

A line is regarded here as the trace of a point moving in space. A plane—as the trace of a line moving in space. A solid—as the trace of a plane moving in space.

Let us imagine a straight line limited by two points, and let us designate this line by the letter a. Let us imagine this line a moving in space in a direction perpendicular to itself and leaving a trace of its movement. When it has traversed a distance equal to its length, the trace left by it will have the form of a square, the sides of which are equal to a line a, i.e. a^2.

Let us imagine this square moving in space in a direction perpendicular to two of its adjoining sides and leaving a trace of its movement. When it has traversed a distance equal to the length of one of the sides of the square, its trace will have the form of a cube, i.e. a^3.

Now if we imagine the movement of a cube in space, what form will the trace left by such a movement, i.e. figure a^4, assume?

Examining the correlations of figures of one, two and three dimensions, i.e. lines, planes and solids, we can deduce the rule that a figure of a higher dimension can be regarded as the trace of the movement of a lower dimension.

On the basis of this rule we may regard figure a^4 as the trace of the movement of a cube in space.

But what is this movement of a cube in space, the trace of which becomes a figure of four dimensions?

If we examine the way in which figures of higher dimensions are constructed by the movement of figures of lower dimensions, we shall discover several common properties and several common laws in these formations.

In fact, when we consider a square as the trace of the movement of a line, we know that all the points of this line have moved in space; when we consider a cube as the trace of the movement of a square, we know that all the points of the square have moved. Moreover, the line moves in a direction perpendicular to itself; the square in a direction perpendicular to two of its dimensions.

Consequently, if we consider the figure a^4 as the trace of the movement of a cube in space, we must remember that all the points of the given cube have moved in space. Moreover, we may deduce from analogy with the above that the cube was moving in space in a direction which is not contained in the cube itself, i.e. a direction perpendicular to its three dimensions. This direction, then, would be the fourth perpendicular unknown to us in our space and in our geometry of three dimensions.

Further, we may determine a line as an infinite number of points ; a square as an infinite number of lines ; a cube as an infinite number of squares. By analogy with this we may determine the figure a^4 as an infinite number of cubes.

Further, looking at the square we see nothing but lines ; looking at the cube we see its surfaces, or possibly even only one of its surfaces.

It is quite possible that the figure a^4 would appear to us as a cube. To put it in a different way, the cube is what we see of the figure a^4.

Further, a point may be determined as a cross-section of a line ; a line as a cross-section of a surface ; a surface as a cross-section of a solid ; a three-dimensional body can therefore be determined as a cross-section of a four-dimensional body.

Generally speaking, in every four-dimensional body we shall see its three-dimensional projection or section. A cube, a sphere, a pyramid, a cone, a cylinder, may be projections or cross-sections of four-dimensional bodies unknown to us.

In 1908 I came across a curious article on the fourth dimension (in Russian) published in the review *Sovremenny Mir.*

It was a letter written by N. A. Morosoff [1] in 1891 to his fellow-prisoners in the fortress of Schlüsselburg. It is of interest chiefly because it contains, in a very picturesque form, an exposition of the fundamental proposition of the method of reasoning about the fourth dimension by means of analogies, which was mentioned above.

The first part of Morosoff's article is very interesting, but in his final conclusions as to what may exist in the domain of the fourth dimension he deviates from the method of analogies and assigns to the fourth dimension the " spirits " which spiritualists evoke in their séances. And then, having denied the existence of spirits, he denies also the objective meaning of the fourth dimension.

[1] N. A. Morosoff, a scientist by education, belonged to the revolutionary parties of the seventies and eighties. He was arrested in connection with the murder of the Emperor Alexander II and spent twenty-three years in prisons, chiefly in the fortress of Schlüsselburg. Liberated in 1905 he wrote several books, one on the Revelation of St. John, another on Alchemy, on Magic, etc., which found fairly numerous readers in the period before the War. It was rather curious that the public liked in Morosoff's books not what he actually wrote, but what he wrote about. His real intentions were very limited and in strict accordance with the scientific ideas of the seventies. He tried to present " mystical subjects " rationally ; for instance, he explained the Revelation as nothing but a description of a thunderstorm. But being a good writer, Morosoff gave a very vivid exposition of his themes, and sometimes he added little-known material. So his books produced a quite unexpected result, and many people became interested in mystical subjects and in mystical literature after reading Morosoff's books. After the revolution, Morosoff joined the Bolsheviks and remained in Russia. Although, as far as is known, he has not taken part in destructive work himself, he has written nothing more and on solemn occasions expresses his official admiration of the Bolshevik régime. (Note to the translation.) P. O.

It is generally supposed that fortress walls do not exist in the fourth dimension, and that was probably the reason why the fourth dimension was one of the favourite subjects of the conversations held at Schlüsselburg by means of tapping.

N. A. Morosoff's letter is an answer to the questions put to him in one of these conversations. He writes :

My dear friends, our short Schlüsselburg summer is nearing its end, and the dark mysterious autumn nights are coming. In these nights, spreading like a black cloak over the roof of our prison and enveloping with impenetrable darkness our little island with its old towers and bastions, it would seem that the shadows of our friends and predecessors who perished here flit invisibly round about these walls, look at us through the windows and enter into mysterious communication with us who still live. And we ourselves, are we not but shadows of what we used to be ? Are we not transformed into some kind of tapping spirits, conversing unseen with one another through the stone walls which divide us, like those that perform at spiritualistic séances.

All day long I have thought of your discussion of to-day about the fourth, the fifth and other dimensions of the space of the universe which are inaccessible to us. With all my power I have tried to imagine at least the fourth dimension of the world, the one in which, as metaphysicians affirm, everything that is under lock and key may suddenly appear open, and in which all confined spaces can be entered by beings able to move not only along our three dimensions, but also along the fourth, to which we are unaccustomed.

You ask me for a scientific examination of the problem. Let us speak first of the world of only two dimensions ; and later we will see whether it will give us the possibility of drawing certain conclusions about different worlds.

Let us take a certain plane—for instance, that which separates the surface of Lake Ladoga which surrounds us, from the atmosphere above it, in this quiet autumn evening. Let us suppose that this plane is a separate world of two dimensions, peopled with its own beings, which can move only on this plane, like the shadows of swallows and sea-gulls flitting in all directions over the smooth surface of the water which surrounds us, but remains for ever hidden from us behind these battlements.

Let us suppose that, having escaped from behind our Schlüsselburg bastions, you went for a bathe in the lake.

As beings of three dimensions you also have the two dimensions which lie on the surface of the water. You will occupy a definite place in the world of shadow beings. All the parts of your body above and below the level of the water will be imperceptible to them, and they will be aware of nothing but your contour, which is outlined by the surface of the lake. Your contour must appear to them as an object of their own world, only very astonishing and miraculous. The first miracle from their point of view will be your sudden appearance in their midst. It can be said with full conviction that the effect you would create would be in no way inferior to the unexpected appearance among ourselves of some ghost from the unknown world. The second miracle would be the surprising changeability of your external form. When you are immersed up to your waist your form will be for them almost elliptical, because only the line on the surface surrounding your waist and impenetrable for them will be perceptible to them. When you begin to swim you will assume in their eyes the outline of a man. When you wade into a shallow place so that the surface on which they live will encircle your legs, you will appear to them transformed into two ring-shaped beings. If, desirous of keeping you in one place, they surround you on all sides, you can step over them and find yourselves free from them in a way quite inconceivable to them. In their eyes you would be all-powerful beings—inhabitants of a higher world, similar to those supernatural beings about whom theologians and metaphysicians tell us.

Now if we suppose that apart from these two worlds, the plane world and the world we live in, there exists a world of four dimensions, superior to ours, it will become clear that in relation to us its inhabitants would be exactly the same as we are in relation to the inhabitants of a plane. They must appear in our midst in the same unexpected way and disappear from our world at their will, moving along the fourth or some other higher dimension.

In a word the analogy, so far, is complete. Further we shall find in the same analogy a complete refutation of all our hypotheses.

If indeed the beings of the four-dimensional world were not purely our invention, their appearance in our midst would be an ordinary, everyday occurrence.

Further Morosoff discusses whether we have any reason to suppose that " supernatural beings " really exist, and he comes to the conclusion that we have no grounds for such a hypothesis unless we are prepared to believe in fairy-tales.

The only indication, worthy of our attention, of the existence of such beings can be found, according to Morosoff, in the teachings of spiritualism. But his own experience in " spiritualism " convinced him that in spite of the strange phenomena that undoubtedly occur at spiritualistic séances, " spirits " take no part in them. So-called " automatic writing ", usually cited as a proof of the co-operation of intelligent forces of another world at these séances, is, according to his observations, a result of thought-reading. Consciously or unconsciously a " medium " " reads " the thoughts of those present and from these thoughts obtains the answers to their questions. Morosoff attended many séances, but never met with a case where there was anything in the answers received which was not known to any of the people present, or where answers were in a language unknown to any present. Therefore, though not doubting the sincerity of the majority of spiritualists, Morosoff concludes that " spirits " have nothing to do with phenomena at séances.

His experience of spiritualism, he says, had finally convinced him many years previously that the phenomena which he assigned to the fourth dimension do not really exist. He says that at such spiritualistic séances answers are given unconsciously by the actual people present and that therefore all suppositions concerning the existence of the fourth dimension are pure imagination.

These conclusions of Morosoff are quite unexpected, and it is difficult to understand how they were arrived at. Nothing can be said against his opinion of spiritualism. The psychic side of spiritualistic phenomena is undoubtedly quite " subjective ". But it is quite incomprehensible why Morosoff sees the " fourth dimension " in spiritualistic phenomena alone, and why, denying the " spirits ", he denies the fourth dimension. This looks like a ready-made solution offered by that official " positivism " to which Morosoff adhered and from which he was unable to depart. His previous arguments led in quite another direction. Besides " spirits " there exist a number of phenomena quite real to us, i.e. of usual and everyday occurrence, but absolutely inexplicable without the help of hypotheses. which would relate these phenomena to the world of the fourth dimension. But we are too accustomed to these phenomena and do not notice their " miraculous character ", do not notice that we live in a world of perpetual miracle, in a world of the mysterious, the inexplicable and, above all, the unmeasurable.

Morosoff describes how miraculous our three-dimensional bodies would seem to the plane-beings, how these beings would not know whence our bodies come and whither they disappear like spirits appearing from an unknown world.

But in reality are we not beings just as fantastic and as changeable in our appearance for any stationary object, a stone or a tree ? Further, do we not possess the properties of " higher beings " for animals ? And are there no phenomena for us, for instance, all the manifestations of *life*, about which we do not know whence they come nor whither they go ; phenomena such as the appearance of a plant from a seed, the birth of living things, and the like ; and further, the phenomena of nature, thunderstorms, rain, spring, autumn, which we can neither explain nor interpret ? Is not each of these phenomena of nature taken separately something of which we can feel only a little, touch only a part, like the blind men in the old Eastern fable who defined an elephant each in his own way : one by its legs, another by its ears, a third by its tail ?

Continuing Morosoff's reasonings concerning the relations between the world of three dimensions and the world of four dimensions, we have no grounds for looking for the latter only in the domain of " spiritualism ".

Let us take a living cell.   It may be exactly equal in length, breadth and height to another, a dead cell.   And still there is something in the living cell which is lacking in the dead one, something we are unable to measure.

We say that it is " vital force ", try to explain the vital force as a kind of motion.   But in reality we do not explain anything by this, but only give a name to a phenomenon which remains inexplicable.

According to some scientific theories vital force must be resolvable into physico-chemical elements, into simpler forces.   But not one of these theories can explain how the one passes into the other and in what relation the one stands to the other.   We are unable to express in a physico-chemical formula the simplest manifestations of life energy.   And as long as we are unable to do so, we have no right, in a strictly logical sense, to regard vital processes as identical with physico-chemical processes.

We may accept philosophical " monism ", but we have no reasons for accepting the physico-chemical monism imposed on us from time to time, which identifies vital and psychic processes with physico-chemical processes.   Our mind may come in an abstract way to the conclusion of the unity of physico-chemical, vital, and psychic processes, but for science, for exact and concrete knowledge, these three classes of phenomena stand quite separate from one another.

For science, three classes of phenomena: mechanical force, vital force and psychic force, pass one into another only partially, and apparently without any fixed or calculable proportions.   Therefore, scientists will be justified in explaining vital and psychic processes as a kind of motion only when they have found means of transforming motion into vital and psychic energy and vice versa, and of calculating such a transformation.   This means that such an affirmation will be possible only when it is known what number of calories contained in a definite quantity of coal is necessary for starting the life of one cell, or how many atmospheres of pressure are necessary for the formation of one thought or one logical deduction.   As long as these are not known, physical, biological and psychic phenomena, as studied by science, take place on different planes.   Their unity can be surmised, but nothing can be affirmed positively.

If one and the same force acts in physico-chemical, vital and psychic processes, it may be supposed that it acts in different spheres only partly contiguous to one another.

If science really possessed knowledge of the unity of at least vital and physico-chemical phenomena, it would be able to create living organisms.   In this expectation there is nothing extravagant.   People construct machines and apparatus which are much more complicated externally than a simple one-cell organism. And yet they are unable to construct such an organism. This means that there is something in a living organism which does not exist in a lifeless machine. A living cell contains something which is lacking in a dead one. And we have every right to call this something equally inexplicable and unmeasurable. And in examining man we have good reasons for putting to ourselves the question : which part is bigger in him, the measurable or the unmeasurable ?

" How can I answer your question " (about the fourth dimension), writes Morosoff in his letter to his fellow prisoners, " when I myself have no dimension in the direction indicated by you ? "

But what real grounds has Morosoff for affirming so definitely that he has not this dimension ?

Can he measure everything in himself ? Two principal functions of man, *life* and *thought*, are in the domain of the unmeasurable.

We know so vaguely and so imperfectly what man really is, and we have in ourselves so much that is enigmatic and incomprehensible from the point of view of the geometry of three dimensions, that we have no reason to deny the fourth dimension in denying " spirits ". On the contrary, we have ample grounds for looking for the fourth dimension precisely in ourselves.

And we have to confess to ourselves clearly and definitely that we do not know in the least what man really is. For us he is an enigma, and we must accept this enigma as such.

The " fourth dimension " promises to explain something in this enigma. Let us try to see what the " fourth dimension " can give us if we approach it with the old methods but without the old prejudices for or against spiritualism. Let us again imagine a world of plane-beings possessing only two dimensions, length and breadth, and inhabiting a flat surface.[1]

Let us imagine, on this surface, living beings having the shape of geometrical figures and capable of moving in two directions.

At the very beginning of the examination of the conditions of life of these flat beings we come at once face to face with a very interesting fact.

These beings will be able to move only in two directions on their plane. They will be unable to rise above this plane or to leave it. In the same way they will be unable to see or feel anything lying outside their plane. If one of these beings rises above the plane, he

[1] In these reasonings about imaginary worlds I shall partly follow Hinton's plan, but this does not mean that I share *all* Hinton's opinions.

will completely pass away from the world of other beings similar to him, will vanish, disappear—no one knows whither.

If we suppose that the organs of vision of these beings are situated on their edges, on their outer lines, then they will not be able to see the world lying outside their plane at all. They will see only lines lying on their plane. They will see each other not as they really are, i.e. in the shape of geometrical figures, but only in the form of lines. In the same way all the objects of their world will also appear to them as lines. And, what is very important, all lines, whether straight, curved, or with angles, or lying at different angles to the line of their edge, will appear to them alike ; they will not be able to see any difference in the lines themselves. But at the same time, the lines will differ for them by strange properties which they will probably call the motion or the vibration of lines.

The centre of a circle will be entirely inaccessible to them. They will be quite unable to see it. In order to reach the centre of a circle a two-dimensional being will have to dig or cut his way through the mass of the flat figure having the thickness of one atom. The process of digging will appear to him as an altering of the line of the circumference.

If a cube is placed on his plane, then this cube will appear to him in the form of the four lines bounding the square touching his plane. Of the whole cube only this square will exist for him. He will be unable even to imagine the rest of the cube. The *cube* will not exist for him.

If several bodies come into contact with his plane, for a plane-being there will exist in each of them only one surface which has come into contact with his plane. This surface, that is, the lines bounding it, will appear to him as an object of his own world.

If through his space, that is, through his plane, there passes a multicoloured cube, the passage of the cube will appear to him as a gradual change in the colour of the lines bounding the square which lies on his plane.

If we suppose that the plane-being is made able to see with his flat side, the one facing our world, it is easy to imagine what a wrong conception of our world he will receive.

The whole universe will appear to him in the form of a plane. It is very probable that he will call this plane æther. Consequently, he will either completely deny all phenomena which take place outside his plane, or regard them as happening on his own plane, in his æther. Unable to explain on his plane all the phenomena observed by him, he may call them miraculous, lying above his understanding, beyond his space, in the " third dimension ".

Having observed that the inexplicable events occur in a certain consecutiveness, in a certain dependence one upon another, and also probably in a dependence on some laws, the plane-being will cease to consider them miraculous and will attempt to explain them by means of more or less complicated hypotheses.

The appearance of the dim idea of another parallel plane will be for a plane-being the first step towards the right understanding of the universe. He will then imagine all the phenomena he is unable to explain on his own plane as occurring on that parallel plane. At this stage of development the whole of our world will appear to him as a plane parallel to his own plane. Neither relief nor perspective will exist for him as yet. A mountain landscape will appear to him as a flat photograph. His conception of the world will certainly be very poor, and full of errors. The big things will be taken for the small, and the small things for the big, and all together, whether near or far, will appear to him equally remote and inaccessible.

Having recognised that there is a world parallel to his plane world, the two-dimensional being will say that of the true nature of the relations between these two worlds he knows nothing.

In the parallel world there will be much that will appear inexplicable for a two-dimensional being. For instance a lever or a couple of wheels on an axle. Their action will appear quite inconceivable to the plane-being, whose conception of laws of motion is limited by motion on a plane. It is quite possible that this phenomenon will be considered supernatural and later will be called, in a more scientific way, " superphysical ".

In studying these superphysical phenomena the plane-being may stumble upon the idea that a lever, or wheels, contain something unmeasurable, but nevertheless existing.

From this there is only one step to the hypothesis of the third dimension. The plane-being will base this hypothesis precisely on inexplicable facts, such as the rotation of wheels. He may ask himself whether the inexplicable may not really be the unmeasurable, and then begin gradually to elucidate for himself the physical laws of three-dimensional space. But he will never be able to prove mathematically the existence of this third dimension, because all his geometrical speculations will proceed only on a plane, on two dimensions, and therefore he will project on a plane the results of his mathematical conclusions, in this way destroying all their meaning.

The plane-being will be able to obtain his first notion of the nature of the third dimension merely by means of logical reasonings and comparisons. This means that in examining the inexplicable that lies in the flat photograph (representing for him our world) the plane-being may arrive at the conclusion that many phenomena are inexplicable for him, because in the objects causing these phenomena there may be a certain *difference* which he does not understand and cannot measure.

Further, he may conclude that a real body must differ in some way from an imaginary one. And having once admitted the hypothesis of the third dimension, he will have to say that the real body, unlike the imaginary body, must possess at least a small third dimension.

In the same way the plane-being may come to the recognition that he must necessarily possess the third dimension.

After arriving at the conclusion that a real body of two dimensions cannot exist, that this is but an imaginary figure, the plane-being will have to say to himself that, since the third dimension exists, he must himself possess this third dimension, because otherwise, having only two dimensions, he would be but an imaginary figure, that is, exist only in somebody's mind.

The plane-being will reason in the following way : " If the third dimension exists, I am either a being of three dimensions or I do not exist in reality but exist only in somebody's imagination ".

In reflecting why he does not see his third dimension the plane-being may come upon the thought that his extension along the third dimension, just like the extension of other bodies along the third dimension, is very small. These reflections may bring the plane-being to the conclusion that for him the question of the third dimension is connected with the problem of small magnitudes.

In investigating the world in a philosophical way the plane-being will from time to time doubt the reality of everything surrounding him and the reality of himself.

He may then think that his conception of the world is wrong and that he does not even see it as it really is. Reasonings about things as they appear and about things as they are may follow from this. The plane-being may think that in the third dimension things must appear as they are, i.e. that he will see in the same things more than he saw in two dimensions.

Verifying all these reasonings from our point of view, that is, from the point of view of beings of three dimensions, we must recognise that all the conclusions of the plane-being are perfectly right and lead him to a right understanding of the world and to the cognition, though theoretical in the beginning, of the third dimension.

We may profit by the experience of the plane-being and try to find whether there is anything in the world towards which we are in the same relation as the plane-being is towards the third dimension.

In examining the physical conditions of the life of man we find in them an almost complete analogy with the conditions of life of the plane-being who begins to be aware of the third dimension.

We shall start by analysing our relation towards the " invisible ".

At first man considers the invisible as miraculous and supernatural. Gradually, with the evolution of knowledge, the idea of the miraculous becomes less and less necessary. Everything within the sphere accessible to observation (and unfortunately far beyond it) is regarded as existing according to certain definite laws, as the result of certain definite causes. But the causes of many phenomena remain hidden, and science is forced to limit itself to a classification of these inexplicable phenomena.

In studying the character and properties of the " inexplicable " in different branches of our knowledge, in physics and chemistry, in biology and in psychology, we can arrive at certain general conclusions concerning the character of the inexplicable. This means that we can formulate the problem as follows . is not the inexplicable a result of something " unmeasurable " for us which exists, first, in those things which, as it appears to us, we can measure fully, and second, in things which, as it appears to us, can have no measurement ?

We can think that this very inexplicability may be the result of the fact that we examine and attempt to explain, within the limits of three dimensions, phenomena that pass into the domain of a higher dimension. To put it differently, are we not in the position of the plane-being trying to explain as happening on a plane phenomena that take place in three-dimensional space ?

There is a great deal that confirms the probability of such a supposition.

It is quite possible that many inexplicable phenomena are inexplicable only because we wish to explain them on our plane, i.e. within our three-dimensional space, while really they occur outside our plane, in the domain of higher dimensions.

Having come to the conclusion that we are surrounded by the world of the unmeasurable, we must admit that, until now, we have had an entirely wrong conception of the objects of our world.

We knew before that we see things and represent them to ourselves not as they really are. Now we may say more definitely that we do not see in things that part of them which is unmeasurable for us, lying in the fourth dimension.

This last conclusion brings us to the idea of the difference between the imaginary and the real.

We saw that the plane-being, having arrived at the idea of the third dimension, must conclude that, if there are three dimensions, a real body of two dimensions cannot exist. A two-dimensional body would be only an imaginary figure, a section of a body of three dimensions or its projection in two-dimensional space.

Admitting the existence of the fourth dimension, we must recognise in the same way that if there are four dimensions, a real body of three dimensions cannot exist. A real body must possess at least a very small extension along the fourth dimension, otherwise it will be only an imaginary figure, the projection of a body of four dimensions in three-dimensional space, like a " cube " drawn on paper.

In this way we must come to the conclusion that there may exist a cube of three dimensions and a cube of four dimensions, and that only the cube of four dimensions will really, actually, exist.

Examining man from this point of view we come to very interesting deductions.

If the fourth dimension exists, one of two things is possible. Either we ourselves possess the fourth dimension, i.e. are beings of four dimensions, or we possess only three dimensions and in that case do not exist at all.

If the fourth dimension exists while we possess only three, it means that we have no real existence, that we exist only in somebody's imagination, and that all our thoughts, feelings and experiences take place in the mind of some other higher being, who visualises us. We are but products of his mind and the whole of our universe is but an artificial world created by his fantasy.

If we do not want to agree with this we must recognise ourselves as beings of four dimensions.

At the same time we must recognise that our own fourth dimension, as well as the fourth dimension of the bodies surrounding us, is known and felt by us only very little and that we only guess its existence from observations of inexplicable phenomena.

Such blindness in relation to the fourth dimension may be caused by the fact that the fourth dimension of our own bodies and other objects of our world is too small and inaccessible to our organs of sense, or to the apparatus which widens the sphere of our observation, exactly in the same way as the molecules of our bodies and many other things are inaccessible to immediate observation. As regards objects possessing a greater extension in the fourth dimension, we feel them at times in certain circumstances, but refuse to recognise them as really existing.

These last considerations give us sufficient grounds for believing that, at least in our physical world, the fourth dimension must refer to the domain of small magnitudes.

The fact that we do not see in things their fourth dimension brings us again to the problem of the imperfection of our perceptions in general.

Even if we leave aside other defects of our perception and regard its activity only in relation to geometry, we shall have to admit that we see everything as very unlike what it really is.

We do not see bodies, we see nothing but surfaces, sides and lines. We never see a cube ; we see only a small part of it, never see it from all sides at once.

From the fourth dimension it must be possible to see the cube from all its sides at once and from within, as though from its centre.

The centre of a sphere is inaccessible to us. To reach it we must cut or dig our way through the mass of the sphere, i.e. act in exactly the same way as the plane-being with regard to the circle. The process of cutting through will in that case appear to us as a gradual change in the surface of the sphere.

The complete analogy of our relation to the sphere with the relation of the plane-being to the circle gives us grounds for thinking that in the fourth dimension, or along the fourth dimension, the centre of the sphere is as easily accessible as is the centre of the circle in the third dimension. In other words, we have a right to suppose that in the fourth dimension it is possible to reach the centre of the sphere from some region unknown to us, along some incomprehensible direction, the sphere itself remaining intact. The latter circumstance would appear to us a kind of miracle, but just as miraculous, to the plane-being, must appear the possibility of reaching the centre of the circle without disturbing the line of its circumference, without breaking up the circle.

Continuing to imagine further the properties of vision or perception in the fourth dimension, we shall have to recognise that not only in a geometrical sense, but also in many other senses, it is possible from the fourth dimension to see in objects of our world much more than we do see.

Prof. Helmholtz once said about our eye that if an optician sent him so badly made an instrument, he would never accept it.

Undoubtedly our eye does not see a great many things which exist. But if in the fourth dimension we see without the aid of such an imperfect instrument, we should be bound to see much more, that is, to see what is invisible for us now and to see everytning without that net of illusions which veils the whole world from us and makes its outward aspect very unlike what it really is.

The question may arise why we should see in the fourth dimension without the aid of eyes, and what this means.

It will be possible to answer these questions definitely only when it is definitely known that the fourth dimension exists and when it is known what it really is. But so far it is possible to consider only what *might* be in the fourth dimension, and therefore there cannot be any final answers to these questions. Vision in the fourth dimension must be effected without the help of eyes. The limits of eyesight are known, and it is known that the human eye can never attain the perfection even of the microscope or telescope. But these instruments with all the increase of the power of vision which they afford do not bring us in the least nearer to the fourth dimension. So it may be concluded that vision in the fourth dimension must be something quite different from ordinary vision. But what can it actually be ? Probably it will be something analogous to the " vision " by which a bird flying over Northern Russia " sees " Egypt, whither it migrates for the winter ; or to the vision of a carrier pigeon which " sees ", hundreds of miles away, its loft, from which it has been taken in a closed basket ; or to the vision of an engineer making the first calculations and first rough drawings of a bridge, who " sees " the bridge and the trains passing over it ; or to the vision of a man who, consulting a time-table, " sees " himself arriving at the station of departure and his train arriving at its destination.

Now, having outlined certain features of the properties which vision in the fourth dimension should possess, we must endeavour to define more exactly what we know of the phenomena of that world.

Again making use of the experience of the two-dimensional being, we must put to ourselves the following question : are all the " phenomena " of our world explicable from the point of view of physical laws ?

There are so many inexplicable phenomena around us that merely by being too familiar with them we cease to notice their inexplicability, and, forgetting it, we begin to classify these phenomena, give them names, include them within different systems and, finally, even begin to deny their inexplicability.

Strictly speaking, all is equally inexplicable. But we are accustomed to regard some orders of phenomena as more explicable and other orders as less explicable. We put the less explicable into a special group, and create out of them a separate world, which is regarded as parallel to the " explicable ".

This refers first of all to the so-called " psychic world ", that is to the world of ideas, images and conceptions, which we regard as parallel to the physical world.

Our relation to the psychic, the difference which exists for us between the physical and the psychic, shows that psychic phenomena should be assigned to the domain of the fourth dimension.[1] In the history of human thought the relation to the psychic is very similar to the relation of the plane-being to the third dimension. Psychic phenomena are inexplicable on the " physical plane ", therefore they are regarded as opposite to the physical. But their unity is vaguely felt, and attempts are constantly made to interpret psychic phenomena as a kind of physical phenomena, or physical phenomena as a kind of psychic phenomena. The division of concepts is recognised to be unsuccessful, but there are no means for their unification.

In the first place the psychic is regarded as quite separate from the body, as a function of the " soul ", unsubjected to any physical laws. The soul lives by itself, and the body by itself, and the one is incommensurable with the other. This is the theory of naïve dualism or spiritualism. The first attempt at an equally naïve monism regards the soul as a direct function of the body. It is then said that " thought is a motion of matter ". Such was the famous formula of Moleschott.

Both views lead into blind alleys. The first, because the obvious interdependence of physiological and psychic processes cannot be disregarded ; the second, because motion still remains motion and thought remains thought.

The first view is analogous to the denial by the two-dimensional being of any physical reality in phenomena which happen outside his plane. The second view is analogous to the attempt to consider as happening on a plane phenomena which happen above it or outside it.

The next step is the hypothesis of a parallel plane on which all the inexplicable phenomena take place. But the theory of parallelism is a very dangerous thing.

The plane-being begins to understand the third dimension when he begins to see that what he considered parallel to his plane may actually be at different distances from it. The idea of relief and perspective will then appear in his mind, and the world and things will take for him the same form as they have for us.

We shall understand more correctly the relation between physical and psychic phenomena when we clearly understand that the psychic is not always parallel to the physical and may be quite independent of it. And parallels which are not always parallel are evidently subject to laws that are incomprehensible to us, to laws of the world of four dimensions.

[1] The expression " psychic " phenomena is used here in its only possible sense of psychological or mental phenomena, that is, those which constitute the subject of psychology. I mention this because in spiritualistic and theosophical literature the word " psychic " is used for the designation of supernormal or superphysical phenomena.

At the present day it is often said : we know nothing about the exact nature of the relations between physical and psychic phenomena ; the only thing we can affirm and which is more or less established is that, for every psychic process, thought or sensation there is a corresponding physiological process, which manifests itself in at least a feeble vibration in nerves and brain fibre and in chemical changes in different tissues. Sensation is defined as the consciousness of a change in the organs of sense. This change is a certain motion which is transmitted into brain centres, but in what way the motion is transformed into a feeling or a thought is not known.

The question arises : is it not possible to suppose that the physical is separated from the psychic by four-dimensional space, i.e. that a physiological process, passing into the domain of the fourth dimension, produces there effects which we call feeling or thought ?

On our plane, i.e. in the world of motion and vibrations accessible to our observations, we are unable to understand or to determine thought, exactly in the same way as the two-dimensional being on his plane is unable to understand or to determine the action of a lever or the motion of a pair of wheels on an axle.

At one time the ideas of E. Mach, expounded chiefly in his book *Analysis of Sensations and Relations of the Physical to the Psychic*, were in great vogue. Mach absolutely denies any difference between the physical and the psychic. In his opinion all the dualism of the usual view of the world resulted from the metaphysical conception of the " thing in itself " and from the conception (an erroneous one according to Mach) of the illusory character of our cognition of things. In Mach's opinion we can perceive nothing wrongly. Things are always exactly what they appear to be. The concept of illusion must disappear entirely. Elements of sensations are physical elements. What are called " bodies " are only complexes of elements of sensations : light sensations, sound sensations, sensations of pressure, etc. Mental images are similar complexes of sensations. There exists no difference between the physical and the psychic ; both the one and the other are built up of the same elements (of sensations). The molecular structure of bodies and the atomic theory are accepted by Mach only as symbols, and he denies them all reality.

In this way, according to Mach's theory, our psychic apparatus builds the physical world. A " thing " is only a complex of sensations.

But in speaking of the theories of Mach it is necessary to remember that the psychic apparatus builds only the " forms " of the world (i.e. makes the world such as we perceive it) out of something else which we shall never attain. The blue of the sky is unreal, the green of the meadows is also unreal; these "colours" belong to the reflected rays. But evidently there is something in the "sky", i.e. in the air of our atmosphere, which makes it appear blue, just as there is something in the grass of the meadow which makes it appear green.

Without this last addition a man might easily have said, on the basis of Mach's ideas : this apple is a complex of my sensations, therefore it only seems to exist, but does not exist in reality.

This would be wrong. The apple exists. And a man can, in a most real way, become convinced of it. But it is not what it appears to be in the three-dimensional world.

The psychic, as opposed to the physical or the three-dimensional, is very similar to what should exist in the fourth dimension, and we have every right to say that thought moves along the fourth dimension.

No obstacles or distances exist for it. It penetrates impenetrable objects, visualises the structure of atoms, calculates the chemical composition of stars, studies life on the bottom of the ocean, the customs and institutions of a race that disappeared tens of thousands of years ago. . . .

No walls, no physical conditions, restrain our fantasy, our imagination.

Did not Morosoff and his comrades fly in their imagination far beyond the bastions of Schlüsselburg ?

Did not Morosoff himself, in his book, *Revelation in Tempest and Thunderstorm*, travel through space and time when, as he was reading Revelations in the Alexeivsky ravelin of the Petropavlovsky Fortress he saw thunder clouds scudding over the Isle of Patmos in the Greek Archipelago, at five o'clock in the afternoon of the 30th September in the year 395 ?

Do we not in sleep live in a fantastic fairy kingdom where everything is capable of transformation, where there is no stability belonging to the physical world, where one man can become another or two men at the same time, where the most improbable things look simple and natural, where events often occur in inverse order, from end to beginning, where we see the symbolical images of ideas and moods, where we talk with the dead, fly in the air, pass through walls, are drowned or burnt, die, and remain alive ?

All this taken together shows us that we have no need to think that the spirits that appear or fail to appear at spiritualistic séances must be the only possible beings of four dimensions. We may have very good reason for saying that we are ourselves beings of four dimensions and are turned towards the third dimension with only one of our sides, i.e. with only a small part of our being. Only this part of us lives in three dimensions, and we are conscious only of this part as our body. The greater part of our being lives in the fourth dimension, but we are unconscious of this greater part of ourselves. Or it would be still more true to say that we live in a four-dimensional world, but are conscious of ourselves only in a three-dimensional world. This means that we live in one kind of conditions, but imagine ourselves to be in another.

The conclusions of psychology bring us to the same idea, but by a different road. Psychology comes, though very slowly, to the recognition of the possibility of awakening our consciousness, i.e. the possibility of a particular state of it, when it sees and feels itself in a real world having nothing in common with this world of things and phenomena—in a world of thoughts, mental images and ideas.

In discussing earlier the properties of the fourth dimension, I mentioned that the tessaract, that is, a^4, may be obtained by the movement of a cube in space, on the condition that all the points of the cube move.

Consequently if we suppose that from each point of the cube there is drawn a line which this movement must follow, the combination of these lines will then form the projection of a body of four dimensions. This body, that is the tessaract, as was found before, can be regarded as an infinite number of cubes growing, as it were, out of the first cube.

Let us see now whether we know of any examples of such motion, which implies the motion of all points of the given cube.

Molecular motion, that is, the motion of minute particles of matter which is increased by heating and lessened by cooling, is the most appropriate example of motion along the fourth dimension, in spite of all the erroneous ideas of physicists with regard to this motion.

In an article entitled " May we hope to see molecules ? " [1] Prof. Goldgammer writes that, according to modern views, molecules are bodies the lineal section of which is something between one millionth and one ten-millionth part of a millimetre. It has been calculated that one milliardth part of a cubic millimetre, that is, one cubic microne, at a temperature of $O°$ C. and at normal pressure contains about 30 million molecules of oxygen. " Molecules move very fast ; thus under normal conditions the majority of molecules of oxygen have the velocity of about 450 metres per second. Molecules do not disperse in all directions instantaneously in spite of their great velocities

[1] In the review *Naoutchnoye Slovo*, February, 1903.

only because they collide every moment with one another and because of this change the direction of their motion. Owing to this the path of a molecule has the aspect of a very entangled zigzag, and a molecule actually ' marks time ', as it were, on one spot."

Leaving aside for the time the entangled zigzag and the theory of colliding molecules (Brownian movement), we must try to find what results are produced by molecular motion in the visible world.

In order to find an example of motion along the fourth dimension we have to find a motion whereby the given body would actually move and not remain in one place (or one state).

Examining all the observable kinds of motion we must admit that the *expansion* and *contraction* of bodies come nearest to the indicated conditions.

Expansion of gases, liquids and solids means that molecules retreat from one another. Contraction of solids, liquids and gases means that the molecules approach one another. The distance between them diminishes. There is space here and there are distances.

Is it not possible that this space lies in the fourth dimension ?

A movement in this space means that all the points of the given geometrical body, that is, all the molecules of the given physical body, move.

The figure resulting from the movement of a cube in space when the cube expands or contracts will have the form of a cube, and we can imagine it as an infinite number of cubes.

Is it right to suppose that the assemblage of lines drawn from every point of a cube, interior as well as exterior, the lines along which the points approach one another or retreat from each other, constitutes the projection of a four-dimensional body ?

In order to answer this it is necessary to determine what these lines are and what this direction is.

These lines connect all the points of the given body with its centre. Consequently the direction of the movement indicated will be from the centre along the radii.

In investigating the paths of the movements of the points (or molecules) of a body in the case of expansion and contraction, we find in them many interesting features.

We cannot see the distance between molecules. We cannot see it in the case of solids, liquids and gases because it is extremely small, and in the case of highly rarefied matter, as for instance that in Crookes tubes, where this distance is probably increased to the proportions perceptible for us or for our apparatus, we cannot see it because the particles themselves, the molecules, are too small to be accessible to our observation. In the above-mentioned article Prof.

Goldgammer states that given certain conditions molecules could be photographed if they could be made luminous. He writes that when the pressure in Crookes tubes is reduced to one-millionth part of an atmosphere one microne will contain only 30 molecules of oxygen. If they were luminous they could be photographed on a screen.

To what extent this photographing is really possible, is another question. For the present argument, a molecule as a real quantity in relation to a physical body can represent a point in its relation to a geometrical body.

All bodies must necessarily consist of molecules; consequently they must possess a certain, though a very small, dimension of inter-molecular space. Without this we cannot conceive a real body, and can conceive only imaginary geometrical bodies. A real body consists of molecules and possesses a certain inter-molecular space.

This means that the difference between a cube of three dimensions, a^3, and a cube of four dimensions, a^4, will be that a cube of four dimensions consists of molecules, whereas a cube of only three dimensions in reality does not exist and is only a projection of a four-dimensional body in three-dimensional space.

In expanding or contracting, that is, in moving along the fourth dimension, if the preceding arguments are admitted, a cube or sphere remains for us all the time a cube or sphere, changing only in size. Hinton quite rightly observed in one of his books that the passing of a cube of higher dimension transversely to our space would appear to us as a change in the properties of the matter of the cube before us. He also says that the idea of the fourth dimension ought to have arisen from observation of a series of progressively growing or diminishing spheres or cubes. This last idea brings him quite near to the right definition of motion in the fourth dimension.

One of the clearest and most comprehensible forms of motion in the fourth dimension in this sense is growth, the principle of which lies in expansion. It is not difficult to explain why it is so. Every motion within the limits of three-dimensional space is at the same time a motion in time. Molecules or points of an expanding cube do not return to their former place on contraction. They trace a certain curve, returning, not to the point of time at which they started, but to another. And if we suppose that generally they do not return, the distance between them and the original point of time will continually increase. Let us imagine the internal motion of a body in the course of which its molecules, having retreated from one another, do not approach one another again, but the distance between them is filled up with new molecules, which in their turn move asunder and make room for new ones. Such an internal motion of a body would be its growth, at least a geometrical scheme of growth. If we compare a little green apple just formed from the ovary with a large red fruit we shall realise that the molecules composing the ovary could not create the apple while moving only in three-dimensional space. They need in addition to this a continuous motion in time, a continuous deviation into the space which lies outside the three-dimensional sphere. The apple is separated from the ovary by time. From this point of view the apple represents three or four months' motion of molecules along the fourth dimension. If we imagine the whole of the way from the ovary to the apple, we shall see the direction of the fourth dimension, that is, the mysterious fourth perpendicular—the line perpendicular to all three perpendiculars of our space and parallel to none of them.

On the whole Hinton stands so near to the correct solution of the problem of the fourth dimension that he sometimes guesses the place of the "fourth dimension" in life, although he cannot determine this place exactly. Thus he says that the symmetry of the structure of living organisms can be explained only by the movement of their particles along the fourth dimension.

Everybody knows, says Hinton,[1] the means of obtaining on paper, images resembling living insects. A few blots of ink are splashed on a piece of paper and the sheet is folded in two. A very complicated symmetrical image is obtained, resembling a fantastic insect. If a whole series of these figures were seen by a man quite unacquainted with the method of their production, then, thinking purely logically, he would have to conclude that they had originated from folding the paper in two, that is to say, that their symmetrically disposed points have been in contact. In the same way, in examining and studying structural forms of organised beings which very strongly resemble the figures on paper obtained by the above-described method, we may conclude that these symmetrical forms of insects, leaves, birds and other animals are produced by means of a process similar to this folding. And we may explain the symmetrical structure of organised beings, if not by folding in two in four-dimensional space, at any rate by a disposition in a manner similar to the folding of the smallest particles from which they are built up.

There exists indeed in nature a very interesting phenomenon, which gives us perfectly correct diagrams of the fourth dimension. It is only necessary to know how to read these diagrams. They are seen in the fantastically varied but always symmetrical shapes of snow-

1 *The Fourth Dimension*, 2nd edition, 1921, pp. 18, 19.

flakes, and also in the designs of the flowers, stars, ferns and lace-work which frost makes on window panes. Drops of water settling from the air on to a cold pane, or on to the ice already formed upon it, begin instantaneously to freeze and expand, leaving traces of their motion along the fourth dimension in the shape of intricate designs. These frost drawings on window panes, as well as the designs of snow-flakes, are figures of the fourth dimension, the mysterious a^4. The motion of a lower figure to obtain a higher one, as imagined in geometry, is here actually realised, and the resulting figure, in effect, represents the trace left by the motion of the lower figure, because the frost preserves all the stages of the expansion of freezing drops of water.

Forms of living bodies, living flowers, living ferns, are created according to the same principles, though in a more complex order. The outline of a tree gradually spreading into branches and twigs is, as it were, a diagram of the fourth dimension, a^4.

FIG. 1.—A diagram of the Fourth Dimension in Nature.

Leafless trees in winter or early spring often present very com-plicated and extraordinarily interesting diagrams of the fourth dimen-sion. We pass them without noticing them because we think that a tree exists in three-dimensional space. Similar wonderful diagrams can be seen in the designs of sea-weeds, flowers, young shoots, certain seeds, etc., etc. Sometimes it is sufficient to magnify them a little in order to see the secrets of the " Great Laboratory " that are hidden from our eye.

Some very remarkable illustrations of the above statements may be found by the reader in Prof. K. Blossfeldt's book on art-forms in nature.[1]

Living organisms, the bodies of animals and human beings, are built on the principles of symmetrical motion. In order to understand these principles let us take a simple schematic example of symmetrical motion. Let us imagine a cube composed of 27 small cubes, and let us imagine this cube as expanding and contracting. During the process of expansion all the 26 cubes lying round the central cube will retreat from it and on contraction will approach it again. For the sake of convenience in reasoning and in order to increase the likeness of the cube to a body consisting of molecules, let us suppose that the cubes have no dimension, that they are nothing but points. In other words, let us take only the centres of the 27 cubes and imagine them connected by lines both with the centre and with each other.

Visualising the expansion of this cube, composed of 27 cubes, we may say that in order to avoid colliding with another cube and hindering its motion, each of these cubes must move away from the centre, that is to say, along the line which connects its centre with the centre of the central cube.

This is the first rule.

In the course of expansion and contraction molecules move along the lines which connect them with the centre.

Further, we see in our cube that the lines connecting the 26 points with the centre are not all equal. The lines drawn to the centre from the centres of the corner cubes are longer than the lines drawn to the centre from the centres of the cubes lying in the middle of the sides of the large cube.

If we suppose that the inter-molecular space is doubled by expansion, then all the lines connecting the 26 points with the centre are at the same time doubled in length. The lines are not equal ; therefore molecules move with unequal speed, some of them faster, and some slower ; those further removed from the centre move faster, those lying nearer the centre move slower.

From this we may deduce a second rule.

The speed of the motion of molecules in the expansion and contraction of a body is proportional to the length of the lines which connect these molecules with the centre.

[1] *Art Forms in Nature*, by Prof. Karl Blossfeldt, with an introduction by Karl Nierendorf (London : A. Zwemmer, 1929).

Observing the expansion of the big cube, we see that the distances between *all* the 27 cubes are increased proportionally to the former distances.

If we designate by the letter *a* lines connecting the 26 points with the centre, and by the letter *b* lines connecting the 26 points with one another, then, having constructed several triangles inside the expanding and contracting cube, we shall see that lines *b* are lengthened proportionally to the lengthening of lines *a*.

From this we deduce a third rule.

In the process of expansion the distance between molecules increases proportionally to the increase of their distance from the centre.

This means therefore that the points that were at an equal distance from the centre will remain at an equal distance from it, and two points that were at an equal distance from a third point will remain at an equal distance from it.

Moreover, if we look upon this motion not from the centre, but from any one of the points, it will appear to us that this point is the centre from which the expansion proceeds, that is to say, it will appear that all the other points retreat from or approach this point, preserving their former relation to it and to each other, while this point itself remains stationary. " The centre is everywhere ! "

The laws of symmetry in the structure of living organisms are based on this last rule. But living organisms are not built by expansion alone. The element of movement in time enters into it. In the course of growth each molecule traces a curve resulting from the combination of two movements, movement in space and movement in time. Growth proceeds in the same direction, along the same lines, as expansion. Therefore the laws of growth must be analogous to the laws of expansion. The conditions of expansion, that is, the *third rule*, ensure the most rigorous symmetry in freely expanding bodies, because if points which were originally at an equal distance from the centre continue always to remain at an equal distance from it, the body will grow symmetrically.

In the figure produced by the ink spread on a sheet of paper folded in two, the symmetry of all the points was obtained because the points on one side came into contact with the points on the other side. To each point on one side there corresponded a point on the other side and, when the paper was folded, these points touched one another. From the third rule formulated above it must follow that between the opposite points of a four-dimensional body there exists some relation, some affinity, which we have not hitherto noticed. To each point there corresponds as it were one or several others linked with it in some way unintelligible to us. That is, this point is unable to move independently; its movement is connected with the movement of other corresponding points, which occupy positions analogous to its own in the expanding and contracting body. And these points are precisely the points opposite to it. It is, as it were, linked with them, linked in the fourth dimension. An expanding body appears to be folded in different ways and this establishes a certain strange connection between its opposite points.

Let us try to examine the way in which the expansion of the simplest

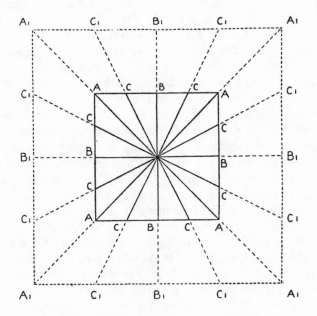

FIG. 2.—Motion from the centre along radii.

figure is effected. We will take this figure not in space even, but on a plane. We will take a square. We will connect the four points at its angles with the centre. Then we will connect with the centre points lying in the middle of the sides, and then points lying half-way between them. The first four points, that is, those lying at the angles, we will call points A; the four points lying in the middle of the sides of the square we will call points B, and finally the points lying also on the sides of the square between A and B (there will be eight of them) we will call points C.

The points A, the points B and the points C lie at different distances from the centre, and therefore on expansion they must move with unequal speed, all the time preserving their relation to the centre. At the same time all the points A are connected among themselves, just as the points B are connected among themselves and as the points C are connected among themselves. Between the points of each group there is a strange inner connection. They must remain at *equal* distances from the centre.

Let us now suppose that the square is expanding, or in other words that all the points, A, B and C retreat from the centre along radii. As long as the expansion of the figure proceeds unhindered, the movement of the points will follow the above-mentioned rules, and the figure will remain a square and preserve a most exact symmetry. But let us suppose that suddenly some obstacle has arisen on the path of the motion of one of the points C, forcing this to stop. In such a case there are two possible alternatives. Either all the other points C will continue to move as if nothing had happened, or they also will stop. If they continue to move, the symmetry of the figure will be broken. If they stop, it will mean a strict observance of the deduction from the third rule, according to which points at an equal distance from the centre must on expansion remain at an equal distance from it. In fact if all the points C^1, obeying the mysterious affinity which exists between them and the point C which met with an obstacle, stop, while points A and B continue to move, then the square will be transformed into a regular, perfectly symmetrical star. It is quite possible that a similar thing happens in the process of the growth of plants and living organisms. Let us take a more complicated figure, in which the centre from which the expansion starts is not a point, but a line, and in which the points retreating from the centre on expansion are disposed on both sides of that line. An analogous expansion will then produce not a star, but something resembling a dentate leaf. If we take this figure as lying in three-dimensional space instead of on a plane and suppose that the centres from which the expansion develops lie not on one but on several axes, we shall obtain on expansion a figure which may resemble a living body with symmetrical limbs, etc.; and if we suppose a movement of the atoms of this figure in time, we shall obtain the " growth " of a living body.

Laws of growth, that is, of motion originating in the centre and proceeding along radii in expansion and contraction, establish a theory which may explain the causes of the symmetrical structure of living bodies.

The definition of states of matter in physics has been becoming more and more conditional. At one time there was an attempt to add to the three generally known states—solid, liquid and gaseous—a fourth, "radiant matter", as the greatly rarefied gases in Crookes tubes were called. Then there exists a theory which considers the colloidal (gelatinous) state of matter as an independent state of matter, different from solid, liquid and gaseous. Organised matter, from the point of view of this theory, is a kind of colloidal matter or is formed from the colloidal matter. The concept of matter in these states was opposed to the concept of energy. Then appeared the electronic theory, in which the concept of matter became very little different from the concept of energy; later came various theories of the structure of the atom, which introduced many new ideas into the concept of matter.

But in this domain more than any other scientific theories differ from ordinary life conceptions. For a direct orientation in the world of phenomena it is necessary for us to distinguish matter from energy, and it is necessary to distinguish the three states of matter—solid, liquid and gaseous. At the same time it must be recognised that even these three states of matter known to us are distinguished by us clearly and indisputably only in their most "classical" forms, like a piece of iron, the water in a river, the air which we breathe. But the transitional forms overlap and are not clear. Therefore very often we do not know exactly when one state passes into the other, cannot draw a definite line of demarcation between the states of matter, cannot say when a solid has been transformed into a liquid, when a liquid has been transformed into gas. We presume that different states of matter depend on a different cohesion of molecules, on the speed and properties of molecular motion, but we distinguish these states only by their external traits, which are very inconstant and often become intermixed.

It can be said definitely that the finer the state of matter the more energetic it is considered to be, that is to say, containing as it were less substance and more motion. If matter is opposed to time, it will be possible to say that each finer state contains more time and less matter than a coarser state.

There is more "time" in a liquid than in a solid; there is more "time" in a gas than in a liquid.

If we accept the possibility of the existence of still finer states of matter, they should be more energetic than those recognised by physics; they should contain, according to the above, more time and less space, still more motion and still less substance.

The logical necessity of energetic states of matter has long been accepted in physics and is proved by very clear reasoning.

... What after all is substance? ... [1] The definition of substance has never been very clear and has become still less clear since the discoveries of modern science. Is it possible, for instance, to define as a substance the mysterious agent to which physicists have recourse for the explanation of phenomena of heat and light? This agent, this medium, this mechanism—call it what you like—nevertheless exists, for it manifests itself in indisputable action. Besides, it is deprived of the qualities without which it is difficult to imagine a substance. It has no weight, and possibly it has no mass ; it does not produce any direct impression on any one of our organs of sense ; in a word it does not possess a single feature which would indicate what was formerly called " material ". On the other hand it is not a spirit, at least nobody has ever thought of calling it that. But does it mean that it is necessary to deny its reality only because it cannot be classified as substance ?

Is it necessary in the same way and for the same reason to deny the reality of the mechanism by means of which gravitation is transmitted into the depths of space with a velocity infinitely greater than the velocity of light,[2] which Laplace considered instantaneous ? The great Newton considered it impossible to do without this agent. He to whom belongs the discovery of universal gravitation wrote to Bentley : " That Gravity should be innate, inherent and essential to Matter, so that one Body may act upon another at a Distance thro' a Vacuum, without the Mediation of anything else, by and through which their Action and Force may be conveyed from one to another, is to me so great an Absurdity, that I believe no Man who has in philosophical Matters a competent Faculty of thinking, can ever fall into it. Gravity must be caused by an Agent acting constantly according to certain Laws ; but whether this Agent be material or immaterial, I have left to the Consideration of my Readers " (3rd letter to Bentley, 25th February, 1692).

The difficulty of allotting a place to these agents is so great that certain physicists, for example Hirn, who has unfolded this idea in his book, *Structure of Celestial Space*, consider it possible to imagine a new class of agents which occupy a position, so to speak, in the middle, between the material and the spiritual order and serve as a great source to the forces of nature. This class of agents, called dynamic by Hirn, from the conception of which he excludes all idea of mass and weight, serves, as it were, to establish relations, to provoke actions over a distance between different parts of matter.

The theory of Hirn's dynamic agents is based upon the following : we could never determine what matter and force really were, but in any case we always considered them opposite to one another, that is to say, we could define matter only as something opposite to force and force as something opposite to matter. But now the old views of matter as something solid and opposite to energy have considerably

---

[1] *Essais sur la philosophie des sciences.* C. de Freycinet (Gauthier Villars & Fils, éditeurs). Paris, 1896, pp. 300-2.
[2] This was written in the nineties of last century.

changed. A physical atom, formerly regarded as indivisible, is now recognised to be complex, composed of electrons. Electrons, however, are not material particles in the usual meaning of the word. They are better defined as moments of manifestation of energy, moments or elements of force. To put it in a different way, electrons, representing the smallest divisions of matter possible, are at the same time the smallest divisions of force. Electrons can be positive or negative. It is possible to think that the difference between matter and force consists simply in different combinations of positive and negative electrons. In one combination they produce on us the impression of matter, in another combination, the impression of force. From this point of view the difference between matter and force, which constitutes so far the basis of our view of nature, does not exist. Matter and force are one and the same thing or, rather, different manifestations of one and the same thing. In any case there is no essential difference between matter and force, and the one must pass into the other. From this point of view matter is nothing but condensed energy. And if it is so, then it is quite natural that degrees of condensation might be different. This theory explains how Hirn was unable to conceive half-material, half-energetic agents. Fine rarefied states of matter must in fact occupy a middle position between matter and force.

In his book *Unknown Forces of Nature*, C. Flammarion wrote : "Matter is not at all what it appears to our senses, to touch or vision. . . . It represents one single whole with energy and is the manifestation of the motion of invisible and imponderable elements. The Universe has a dynamic character. Guillaume de Fontenay gives the following explanation of the dynamic theory. In his opinion matter is in no way the inert substance it is usually considered to be."

Let us take a carriage wheel and place it horizontally on the axle. The wheel is not moving. Let us take a rubber ball and make it fall between the spokes. Now let us make the wheel move slightly. The ball will fairly often hit the spokes and rebound. If we increase the rotation of the wheel the ball will not pass through it at all ; the wheel will become for it a kind of impenetrable disc. We may make a similar experiment placing the wheel vertically and pushing a rod through it. A bicycle wheel will serve the purpose well, as its spokes are thin. When the wheel is stationary, the rod will pass through it nine times out of ten. When in motion the wheel will repel the rod more and more often. When the speed of its motion is increased it will become impenetrable, and all efforts at piercing it will strike as against steel armour.[1]

---

[1] Camille Flammarion, *Les forces naturelles inconnues*. Paris, 1927 (E. Flammarion, éditeur), p. 568.

Now having examined in the world surrounding us all that answers to the physical conditions of a higher dimensional space, we may put the question more definitely : what is the fourth dimension ?

We have seen that it is impossible to prove its existence mathematically or to determine its properties and above all to define its position in relation to our world. Mathematics admits only the possibility of the existence of higher dimensions.

At the very beginning, when defining the idea of the fourth dimension, I pointed out that if it existed, it would mean that besides the three perpendiculars known to us there must exist a fourth. And this in its turn would mean that from any point of our space a line can be traced in a direction unknown and unknowable for us, and further that quite close, side by side with us, but in an unknown direction, there lies some other space which we are unable to see and into which we cannot pass.

I explained later why we are unable to see this space and I determined that it must lie not side by side with us in an unknown direction, but inside us, inside the objects of our world, inside our atmosphere, inside our space. However, this is not the solution of the whole problem, although it is a necessary stage on the way to this solution, because the fourth dimension *is not only inside us*, but we ourselves are inside it, that is, in the space of four dimensions.

I mentioned before that " spiritualists " and " occultists " of different schools often use the expression " fourth dimension " in their literature, assigning to the fourth dimension all phenomena of the " astral sphere ".

The " astral sphere " of the occultists which permeates our space is an attempt to find a place for phenomena which do not fit into our space. And consequently it is to a certain extent that continuation of our world inwards which we require.

The " astral sphere " from an ordinary point of view may be defined as the *subjective world*, projected outside us and taken for the *objective world*. If anybody actually succeeded in establishing the objective existence of even a portion of what is called " astral ", it would be the world of the fourth dimension.

But the very concept of the " astral sphere " or " astral matter " has changed many times in occult teachings.

On the whole, if we take the views of " occultists " of different schools on nature, we shall see that they are based upon the recognition of the possibility of studying conditions of existence other than our physical ones, and of using the knowledge of these other conditions of existence for the purpose of influencing our physical conditions. " Occult " theories generally start from the recognition of one basic substance, the knowledge of which provides *a key* to the knowledge of the mysteries of nature. But the concept of this substance is not definite. Sometimes it is understood as a *principle*, as a *condition of existence*, and sometimes as *matter*. In the first instance the basic substance contains in itself the roots and causes of things and events ; in the second instance the basic substance is the primary matter from which everything else is obtained. The first concept is of course much more subtle and is the result of more elaborate philosophical thought. The second concept is more crude and is in most cases a sign of the decline of thought, a sign of an ignorant handling of difficult and profound ideas.

Philosopher-alchemists called this fundamental substance " Spiritus Mundi "—the spirit of the world. But alchemists—*seekers after gold*—considered it possible to put the spirit of the world into a crucible and subject it to chemical manipulations.

This should be kept in mind in order to understand the " astral hypotheses " of modern theosophists and occultists. Saint-Martin and, later, Eliphas Lévi still understood the " astral light " as a *principle*, as conditions of existence other than our physical conditions. But in the case of modern spiritualists and theosophists " astral light " has been transformed into " astral matter ", which can be *seen* and even photographed. The theory of " astral matter " is based on the hypothesis of " fine states of matter ". The hypothesis of fine states of matter was still possible in the last decades of the old physics, but it is difficult to find a place for it in modern physico-chemical thought. On the other hand, modern physiology deviates further and further from physico-mechanical explanations of vital processes and comes to the recognition of the enormous influence of *traces of matter*, that is, of imponderable and chemically indefinable matters, which are nevertheless clearly seen by the results of their presence, such as " hormones ", " vitamines ", " internal secretions " and so on.

Therefore, in spite of the fact that the hypothesis of fine states of matter does not stand in any relation whatever to new physics I shall attempt here to give a short exposition of the " astral theory ".

According to this theory particles resulting from the division of physical atoms produce a kind of special fine matter—" astral matter " —unsubjected to the action of the majority of physical forces, but subjected to the action of forces not affecting physical matter. Thus this " astral matter " is subjected to the action of psychic energy, will, feelings and desires, which are real forces in the astral sphere. This means that man's will, and also his sense reactions and emotional impulses, act upon " astral matter " just as physical energy acts upon physical bodies.

Further, the transformation into the astral state of physical matter composing visible bodies and objects is recognised as possible. This is *dematerialisation*, that is, from the physical point of view, a complete disappearance of physical objects no one knows where without trace or remains. Also, the reverse process, that is, the transformation of astral matter into the physical state or into physical matter is recognised as possible. This is *materialisation*, that is, the appearance of things, objects and even living beings from no one knows where.

Moreover, it is recognised as possible that matter which enters into the composition of a physical body, after having been transformed into the astral state, may "return" to the physical state in another form. Thus one metal, having been transformed into the astral state, may "return" in the form of another metal. In this way alchemical processes are explained by the temporary transference of some body, most often some metal, into an astral state where matter is subject to the action of will (or of spirits) and may change entirely under the influence of this will and reappear in the physical world *as another metal*; thus iron can change into gold. It is recognised as possible to accomplish this transformation of matter from one state into another and the transformation of one body into another by means of mental influence, assisted by certain rituals, etc. Further it is considered possible to see in the astral sphere events which have not yet happened in the physical sphere, but which must happen and must influence both the past and the future.

All this taken together makes up the content of what is called magic.

Magic, in the usual understanding of this word, means the capacity to accomplish what cannot be accomplished by ordinary physical means. Such would be, for instance, the power to influence psychically people and objects at a distance, to see people's actions and to know their thoughts, to make them disappear from our world and appear in unexpected places; the capacity to change one's appearance and even one's physical nature, to transfer oneself in some inconceivable way to great distances, to pass through walls, etc.

"Occultists" explain all such acts by the knowledge of the properties of the "astral sphere" possessed by magicians and their ability to act mentally upon astral matter and through it upon physical matter. Certain kinds of "sorcery" can be explained by the imparting of special properties to inanimate objects. This is attained by means of influencing psychically their "astral matter", by a special kind of psychic magnetisation of them; in this way magicians could impart to objects any properties they chose, make them execute their will, bring good or evil to other people, warn them against impending disasters, give force or take force away. To such magical practices belongs, for instance, the "blessing of water", which has now become nothing but a rite in Christian and Buddhist religious services. Originally it was an operation undertaken for the purpose of saturating water psychically with certain radiations or emanations with the aim of endowing it with the desired qualities, curative or other.

In theosophical and modern occult literature there are many very picturesque descriptions of the astral sphere. But no proofs of the objective existence of the astral sphere are anywhere given.

" Spiritualistic " proofs, that is, phenomena at séances, or " medium-istic " phenomena in general, " communications ", etc., ascribed to spirits, that is, to disincarnated souls, are in no sense proofs, because all these phenomena can be explained much more simply. In the chapter on dreams I point out the possible meaning of spiritualistic phenomena as the results of impersonation. Theosophical explanations based upon " clairvoyance " require first of all proof of the existence of " clairvoyance ", which remains unproved in spite of the number of books in which the authors have described what they attained or what they found by means of clairvoyance.

It is not generally known that in France there exists a prize, established many years ago, which offers a considerable sum of money to anybody who would read a letter in a closed envelope. The prize remains unclaimed.

Both the spiritualistic and the theosophical theories suffer from one common defect which explains why " astral " hypotheses remain always the same and receive no proofs. " Space " and " time " are taken both in spiritualistic and in theosophical astral theories in exactly the same way as in the old physics, that is, separately from one another. " Disincarnated spirits " or " astral beings " or thought forms are taken *spatially* as bodies of the fourth dimension, but *in time* as physical bodies. In other words they remain in the same time conditions as physical bodies. And it is precisely this that is impossible. If " fine states of matter " produce bodies of different spatial existence, these bodies must have a different time existence. But this idea does not enter into theosophical or spiritualistic thought.

In this chapter there has been collected only the historical material relating to the study of the " fourth dimension ", or rather that part of the historical material which brings one nearer to the solution of the problem or at least to its more exact formulation.

In this book, in the chapter " A New Model of the Universe ", I show how the problems of " space-time " are connected with the problems of the structure of matter, and consequently the structure of the world, and how they lead to a right understanding of the *real* world, avoiding a whole series of unnecessary hypotheses, both pseudo-occult and pseudo-scientific.

1908–1929.

CHAPTER III
EXPERIMENTAL MYSTICISM

Magic and mysticism—Basic propositions—Methods of magic operations—Purpose of my experiments—The beginning of the experiments—First realisations—Sensation of duality—An unknown world—Absence of separateness—Infinite number of new impressions—Change in relation between subjective and objective—World of complicated mathematical relations—Formation of a design—Attempts to express visual impressions in words—Attempts to converse during experiments—Feeling of lengthening of time —Attempts to make notes during experiments—Connection between breathing and heart-beat—Moment of second transition—" Voices " of transitional state—Rôle of imagination in transitional states—The new world beyond the second threshold—Infinity —Mental world " Arupa "—Realisation of danger—Emotionality of experiences— The number three—Another world within the usual world—All things connected— Old houses—A horse in the Nevsky—Attempts to formulate—" Thinking in other categories "—Coming into contact with oneself—" I " and " he "—" Ash-tray "— " Everything is alive "—Symbol of the world—Moving signs of things or symbols— Possibility of influencing another man's fate—Consciousnesses of physical body— Attempts to see at a distance—Two cases of strengthening capacity of perception— Fundamental error of our thinking—Non-existent ideas—Idea of triad—Idea of " I " —Ordinary sensation of " I "—Three different cognitions—Personal interest—Magic —Cognition based on calculation—Sensations connected with death—" Long body of life "—Responsibility for events in the life of another man—Connection with the past and with other people—Two aspects of the phenomena of the world—Return to ordinary state—Dead world in place of living world—Results of experiments.

IN 1910 and 1911, as a result of a fairly complete acquaintance with existing literature on " theosophy " and " occultism " and also with the not very numerous scientific investigations of phenomena of *witchcraft, sorcery, magic*, etc., I came to certain definite conclusions, which I was able to formulate in the following propositions :

1. All manifestations of any unusual and supernormal forces of man, both internal and external, should be divided into two main categories—*magic* and *mysticism*. Definition of these concepts presents great difficulties, because, first, in general as well as in special literature both terms are very often used in an entirely wrong sense ; second, there remains much that is unexplained in respect both of magic and of mysticism taken separately ; and third, the relation of magic and mysticism to one another remains similarly unexplained.

2. Having ascertained the difficulty of exact definition I decided to accept an approximate definition.

I called *magic* all cases of intensified doing or of concrete knowing through other than ordinary means, and I divided magic into *objective*, i.e. with real results, and *subjective*, i.e. with imaginary results. And I called *mysticism* all cases of intensified feeling and abstract knowing.

141

I called objective magic intensified doing and concrete knowing. " Intensified doing " means in this case the *real* possibility of influencing things, events and people without the aid of ordinary means, at a distance, through walls, or in time, that is, either in the past or in the future, and further, the possibility of influencing the " astral " world, if such a world exists, that is, the souls of the dead, " elementals ", forces unknown to us, whether good or evil. Concrete knowing includes clairvoyance in space and time, " telepathy ", thought-reading, psychometry, seeing " spirits ", " thought-forms ", " auras " and the like, again if all these exist.

I called subjective magic all cases of *imaginary* doing and *imaginary* knowing ; in this are included artificially evoked hallucinations, dreams taken as reality, the reading of *one's own* thoughts taken as communications, the semi-intentional creation of astral visions, " Akashic records " and similar miracles.

Mysticism in its nature is subjective. I did not therefore put objective mysticism into a special group. I nevertheless found it possible sometimes to call *subjective mysticism* the false mystical states or pseudo-mystical states which are not connected with *intensified* feeling, but come near hysteria and pseudo-magic ; in other words religious visions or religious dreams in concrete forms, that is, all that in Orthodox literature is called " beauty ".

3. The existence of *objective magic* cannot be considered established. Scientific thought has long denied it and recognised only subjective magic, that is, a kind of self-hypnosis, or hypnosis. In recent times certain admissions are met with in scientific literature or in literature that is intended to be scientific, for instance in the direction of " spiritualism ". But these latest admissions are as unreliable as previous denials. " Theosophical " and " occult " thought recognises the possibility of objective magic, but in some cases evidently confuses it with mysticism, and in other cases opposes it to mysticism as a phenomenon *useless* and *immoral*, or at any rate *dangerous*, both for the man who practises " magic " and for other people, and even for the whole of humanity. But all this is affirmed though satisfactory proofs of the real existence and possibility of objective magic are absent.

4. Of all the unusual states of man there can be regarded as fully established only mystical states of consciousness and certain phenomena of subjective magic, these latter being almost all confined to the artificial creation of the visions desired.

5. All the established facts relating to the manifestations of any unusual forces of man, both in the domain of magic, even though

subjective, and in the domain of mysticism, are connected with greatly intensified emotional states of a particular kind and never occur without them.

6. The greater part of the religious practice of all religions, and also various magic rituals, ceremonies and the like, have as their aim the creation of these emotional states, to which, according to the original intention, either " magical " or " mystical " powers are ascribed.

7. In many cases of deliberate creation of mystical states or production of magical phenomena the use of narcotics can be traced. In all religions of ancient origin, even in their modern form, there still survives the use of incense, perfumes, unguents, which may primarily have been connected with the use of drugs affecting the emotional and intellectual functions. of man. As can be traced, drugs of that kind were very largely used in the ancient Mysteries. Many authors have pointed out the rôle of the sacred drink which was given to candidates for initiation, for instance in the Eleusinian Mysteries, and which may have had a very real and not in the least a symbolical meaning. The legendary sacred drink, the " Soma ", which plays a very important part in Indian mythology and in the description of different kinds of mystical ceremonies, may have actually existed as a drink which brought people into a definite, desired state. In all descriptions of witchcraft and sorcery in all countries and among all peoples, the use of narcotics is invariably mentioned. The witches' ointments which served for flying to the Sabbath, different kinds of enchanted and magical drinks, were prepared either from plants possessing stimulant, intoxicant and narcotic properties, or from organic extracts of the same character, or from those vegetable or animal substances to which these properties were ascribed. It is known that in these cases as well as in all kinds of sorcery, belladonna, datura, extracts of poppy (opium), and, especially, of hemp (hashish) were used. All this can be traced and verified, and leaves no doubts as to its meaning. The African wizards, with regard to whom it is possible to find very interesting descriptions in the accounts of modern explorers, use hashish very largely. Siberian Shamans in order to produce in themselves a particular excited state, in which they can foretell the future (real or imaginary), or influence those about them, make use of poisonous mushrooms (*crimson fly-agaric*).

Interesting observations on the meaning of mystical states of consciousness and on the part which may be played by narcotics in the creation of mystical states can be found in Prof. James' book *Varieties of Religious Experience* (New York, 1902).

Various exercises of Yogis : breathing exercises, unusual postures, movements, " sacred dances ", etc., have the same object, that is, the creation of mystical states of consciousness. But these methods are still very little known.

In examining the above propositions from the point of view of different methods I came to the conclusion that a new experimental verification of the possible results of the application of these methods was necessary, and I decided to start a series of experiments.

The following is a description of the effects I obtained by applying to myself certain methods, the details of which I had partly found in the literature on these subjects, and partly derived from all that has been set forth above.

I do not describe the actual methods I used. First, because it is not the methods but the results that matter, and second, because the description of methods would divert attention from the facts I intend to examine.

I hope some time later to return specially to these " methods ".

My task, as I formulated it to myself at the beginning of my experiments, was to elucidate the questions of the relation of subjective magic to objective magic and then of the relation of objective and subjective magic, taken together, to mysticism.

All this took the shape of three questions :

1. Can the real existence of objective magic be recognised ?
2. Does objective magic exist without subjective ?
3. Does objective magic exist without mysticism ?

Mysticism as such interested me less. However, I said to myself that if we could find a means of deliberately changing our state of consciousness, while at the same time preserving the faculty of self-observation, that would give us completely new material for self-study. We always see ourselves from one and the same angle. If what I supposed should prove to be right, it would mean that we could see ourselves from entirely new and unexpected angles.

The very first experiments showed me the difficulty of the task I had set myself and partly explained to me the failure of many experiments which had been tried by others before me.

A change in the state of consciousness as a result of my experiments began to take place very soon, much more quickly and easily than I thought. But the chief difficulty was that the new state of consciousness which was obtained gave at once so much that was new and unexpected, and these new and unexpected experiences came upon me and flashed by so quickly, that I could not find words, could not find forms of speech, could not find concepts, which would enable me to remember what had occurred even for myself, still less to convey it to anyone else.

The first new psychic sensation which appeared was a sensation of strange duality in myself. Such sensations occur, for instance, in moments of great danger or, in general, under the stress of strong emotions, when a man does or says something almost automatically and at the same time observes himself. This sensation of duality was the first new psychic sensation which appeared in my experiments, and it usually remained throughout even the strangest and most fantastic experiences. There was always a certain point which observed. Unfortunately it could not always remember what it had observed.

The changes in psychic states, this " duality of personality " that occurred, and many other things which were connected with it, usually began about twenty minutes after the beginning of the experiment. When this change came I found myself in a world entirely new and entirely unknown to me, which had nothing in common with the world in which we live, still less with the world which we assume to be the continuation of our world in the direction of the unknown.

That was one of the first strange sensations which struck me. Whether we confess it to ourselves or not, we have a certain conception of the unknowable and of the unknown, or, to be more exact, a certain expectation of it. We expect to see a world which is strange but which consists on the whole of the same kind of phenomena we are accustomed to, or which exists according to the same laws, or has at least something in common with the world we know. We cannot imagine anything *new*, just as we should not be able to imagine an entirely new animal which does not resemble in any way any of the animals we know.

And in this case I saw from the very beginning that all that we half-consciously construct with regard to the unknown is completely and utterly wrong. The unknown is unlike anything that we can suppose about it. The complete unexpectedness of everything that is met with in these experiences, from great to small, makes the description of them difficult. First of all, everything is unified, everything is linked together, everything is explained by something else and in its turn explains another thing. There is nothing separate, that is, nothing that can be named or described *separately*. In order to describe the first impressions, the first sensations, it is necessary to describe *all* at once. The new world with which one comes into contact has no sides, so that it is impossible to describe first one side and then the other. All of it is visible at once at every point; but

how in fact to describe anything in these conditions—that question
I could not answer.

I understood why all descriptions of mystical experiences are so
poor, so monotonous and obviously invented. A man becomes lost
amidst the infinite number of totally new impressions, for the expression of which he has neither words nor forms. When he wishes to
express or convey them to somebody else he involuntarily uses words
which in his ordinary language correspond to the greatest, the most
powerful, the most unusual and the most extraordinary, though these
words do not in the least correspond to what he sees, learns or experiences. The fact is that he has no other words. But in most cases
the man is not even aware of this substitution because his experiences are preserved in his memory as they actually were only for a
few moments. Very soon they fade, grow flat, are replaced by the
words which were hurriedly and accidentally attached to them to keep
them in memory. Very soon nothing remains but these words.
This explains why a man who has had mystical experiences uses,
for expressing and transmitting them, those forms of images, words
and speech which are best known to him, which he is accustomed
to use most often and which are the most typical and characteristic
for him. In this way it may easily happen that different people
describe and convey an entirely identical experience quite differently.
A religious man will make use of the usual clichés of his religion.
He will speak of the Crucified Jesus, of the Virgin Mary, of the Holy
Trinity, and so on. A philosopher will try to render his experiences
in the language of the metaphysics to which he is accustomed. For
instance he will speak of " categories " or of " monads ", or of
" transcendental qualities ", or something of the sort. A theosophist
will speak of the " astral " world, of " thought forms " and of
" Teachers ". A spiritualist will speak of the spirits of the dead and
of communication with them. A poet will speak of his experiences
in the language of fairy-tales or ancient myths, or by describing them
as sensations of love, rapture, ecstasy.

My personal impression was that in the world with which I came
into contact there was nothing resembling any of the descriptions
which I had read or heard of before.

One of the first impressions which astonished me was that in
this world there was absolutely nothing in any way resembling the
theosophical or spiritualistic " astral world ". I say " astonished ",
not because I actually believed in this astral world, but because
probably I had unconsciously thought about the unknown in forms
of the astral world. As a matter of fact, at that time I was to
a certain extent under the influence of theosophical literature, in so
far, at any rate, as refers to nomenclature. To put it more correctly,

I evidently thought, without formulating it quite clearly, that something must lie behind those perfectly concrete descriptions of the invisible world which are to be found in theosophical books. So that at first it was difficult for me to admit that the whole astral world that was described in such detail by different authors did not exist at all. Later, I found that many other things also did not exist.

I will try to describe in short what I met with in this strange world in which I saw myself.

What I first noticed, simultaneously with the " division of myself into two ", was that the relation between the objective and the subjective was broken, entirely altered, and took certain forms incomprehensible to us. But " objective " and " subjective " are only words. I do not wish to hide behind these words, but I wish to describe as exactly as possible what I really felt. For this purpose I must explain what it is that I call " objective " and " subjective ". My hand, the pen with which I write, the table, these are objective phenomena. My thoughts, my mental images, the pictures of my imagination, these are subjective phenomena. The world is divided for us along these lines when we are in our ordinary state of consciousness, and all our ordinary orientation works along the lines of this division. In the new state all this was completely upset. First of all we are accustomed to the constancy of the relation between the subjective and the objective—what is objective is always objective, what is subjective is always subjective. Here I saw that the objective and the subjective could change places. The one could become the other. It is very difficult to express this. The habitual mistrust of the subjective disappeared ; every thought, every feeling, every image, was immediately objectified in real substantial forms which differed in no way from the forms of objective phenomena ; and at the same time objective phenomena somehow disappeared, lost all reality, appeared entirely subjective, fictitious, invented, having no real existence.

This was the first experience. Further, in trying to describe this strange world in which I saw myself, I must say that it resembled more than anything a world of *very complicated mathematical relations*.

Imagine a world in which all relations of quantities, from the simplest to the most complicated, have a form.

Certainly it is easy to say " imagine such a world ".

I understand perfectly well that to " imagine " it is impossible. Yet at the same time what I am saying is the closest approximation to the truth which can be made.

" A world of mathematical relations "—this means a world in

which everything is connected, in which nothing exists separately and in which at the same time the relations between things have a real existence apart from the things themselves ; or, possibly, " things " do not even exist and only " relations " exist.

I am not deceiving myself, and I realise that my descriptions are very poor and will probably not convey what I myself remember. But I remember seeing mathematical laws in operation, and the world as the result of the operation of these laws. Thus the process of the world's creation, when I thought of it, appeared to me under the aspect of the differentiation of some very simple basic principles or basic quantities. This differentiation always proceeded before my eyes in certain forms, sometimes for instance taking the form of a very complicated design developing out of a very simple basic *motif*, which was continually repeated and entered into every combination throughout the design. Thus the whole of the design consisted of nothing but combinations and repetitions of the basic *motif* and could at any point, so to speak, be resolved into its component elements. Sometimes it was music, which began similarly with some very simple sounds and gradually passed into complicated harmonious combinations expressed in visible forms, resembling the design which I have just described, or completely merging into it. The music and the design made a single whole : the one as it were expressed the other.

Throughout the strangest experiences I always felt that nothing of them would remain when I returned to my ordinary state. I understood that in order to remember what I had seen and felt it had all to be translated into words. But for many things there were no words, while other things passed before me so quickly that I had no time to connect them with any words. Even at the time, in the middle of these experiences, I felt that what I was remembering was only an insignificant part of what had passed through my consciousness. I continually said to myself : " I must at least remember that *this is*, that *this was*, and that this is the only reality, while everything else in comparison with it is not real at all."

I tried my experiments under the most varied conditions and in the most varied surroundings. Gradually I became convinced that it was best to be alone. Verification of the experiments, that is, observation by another person, or the recording of the experiences at the very moment they took place, was quite impossible. In any case I never obtained any results in this way.

When I tried having someone near me during these experiments, I found that no kind of conversation could be carried on. I began to say something, but between the first and second words of my sentence such an enormous number of ideas occurred to me and passed before me, that the two words were so widely separated as to make it impossible to find any connection between them. And the third word I usually forgot before it was pronounced, and in trying to recall it I found a million new ideas, but completely forgot where I had begun. I remember for instance the beginning of a sentence : "I said yesterday " . . .

No sooner had I pronounced the word "I" than a number of ideas began to turn in my head about the meaning of the word, in a philosophical, in a psychological and in every other sense. This was all so important, so new and profound, that when I pronounced the word " said ", I could not understand in the least what I meant by it. Tearing myself away with difficulty from the first cycle of thoughts about "I", I passed to the idea " said ", and immediately found in it an infinite content. The idea of speech, the possibility of expressing thoughts in words, the past tense of the verb, each of these ideas produced an explosion of thoughts, conjectures, comparisons and associations. Thus, when I pronounced the word " yesterday " I was already quite unable to understand why I had said it. But it in its turn immediately dragged me into the depths of the problems of time, of past, present and future, and before me such possibilities of approach to these problems began to open up that my breath was taken away.

It was precisely these attempts at conversation, made in these strange states of consciousness, which gave me the sensation of change in time which is described by almost everyone who has made experiments like mine. This is a feeling of the extraordinary lengthening of time, in which seconds seem to be years or decades.

Nevertheless, the usual feeling of time did not disappear; only together with it or within it there appeared as it were another feeling of time, and two moments of ordinary time, like two words of my sentence, could be separated by long periods of another time.

I remember how much I was struck by this sensation the first time I had it. My companion was saying something. Between each sound of his voice, between each movement of his lips, long periods of time passed. When he had finished a short sentence, the meaning of which did not reach me at all, I felt I had lived through so much during that time that we should never be able to understand one another again, that I had gone too far from him. It seemed to me that we were still able to speak and to a certain extent understand one another at the beginning of this sentence, but by the end it had become quite impossible, because there were no means of conveying to him all that I had lived through in between.

Attempts at writing also gave no results, except on two or three occasions, when short formulations of my thoughts, written down during the experiment, enabled me afterwards to understand and decipher something out of a series of confused and indefinite recollections. But generally everything ended with the first word. It was very rarely that I went further. Sometimes I succeeded in writing down a sentence, but usually as I was finishing it I did not remember and did not understand what it meant or why I had written it, nor could I remember this afterwards.

I will try to describe consecutively how my experiments proceeded.

I omit the physiological phenomena which preceded the change in my psychic state. I will mention only that the pulsation now quickened, reaching a very high rate, now slowed down.

In this connection I several times observed a very interesting phenomenon.

In the ordinary state *intentional* slowing down or acceleration of the breathing equally produces accelerated beating of the heart. But in this case, entirely without intention on my part, there was established between the breathing and the beating of the heart a connection which ordinarily does not exist; namely, by accelerating the breathing I accelerated the beating of the heart, by slowing down the breathing I slowed down the beating of the heart. I felt that behind this new capacity lay very great possibilities. I tried therefore not to interfere with the work of the organism but to let things follow their natural course.

Left to itself, the pulsation was intensified and was gradually felt in various parts of the body as though gaining more and more ground, and at the same time it became gradually balanced until at last it began to be felt throughout the body simultaneously and after that continued as *one beat*.

This synchronised pulsation went on quickening, and suddenly a shock was felt through the whole body as though a spring clicked, and at the same instant something opened in me. Everything suddenly changed, there began something strange, new, entirely unlike anything that occurs in life. This I called the first threshold.

There was in this new state a great deal that was incomprehensible and unexpected, chiefly in the sense of still greater confusion of objective and subjective; and there were also other new phenomena of which I will now speak. But this state was not yet complete. It should more properly be called the transitional state. In many cases my experiments did not take me further than this state. Sometimes, however, it happened that this state deepened and widened as though I was gradually plunged in light. After that there came a moment of yet another transition, again a kind of shock throughout the body. And only after this began the most interesting state which I attained in my experiments.

The " transitional state " contained almost all the elements of this state, but at the same time it lacked something most important and essential. The " transitional state " did not differ much in its essence from dreams, especially from dreams in the " half-dream state ", though it had its own very characteristic forms. And the " transitional state " might perhaps have taken me in by a certain sensation of the miraculous that was connected with it, if I had not been able to adopt a sufficiently critical attitude towards it, based chiefly on my earlier experiments in the study of dreams.

In the " transitional state ", which, as I learned very soon, was entirely subjective, I usually began almost at once to hear " voices ". These " voices " were a characteristic feature of the " transitional state ".

The voices spoke to me and often said very strange things which seemed to have a quality of trick in them. Sometimes in the first moments I was excited by what I heard in this way, particularly as it answered certain vague and unformulated expectations that I had. Sometimes I heard music which evoked in me very varied and powerful emotions.

But strangely enough I felt from the first day a distrust of these states. They contained too many promises, too many things I wanted to have. The voices spoke about every possible kind of thing. They warned me. They proved and explained to me everything in the world, but somehow they did it too simply. I began to ask myself whether I might not myself have invented all that they said, whether it might not be my own imagination, that unconscious imagination which creates our dreams, in which we can see people, talk to them, hear their voices, receive advice from them, etc. After thinking in this way I had to say to myself that the voices told me nothing that I could not have thought myself.

At the same time what came in this way was often very similar to the " communications " received at mediumistic séances, or by means of automatic writing. The voices often gave themselves different names, said various flattering things to me and undertook to answer all kinds of questions. Sometimes I had long conversations with these voices.

Once I asked a question referring to alchemy. I cannot now remember the exact question, but I think it was something either about the different denominations of the four elements: fire, water, air and earth; or about the relation of the four elements to one another. I put the question in connection with what I was reading at the time.

In answer to this question a voice which called itself by a well-known name told me that the answer to my question would be found in a certain book. When I said that I had not got this book the voice told me that I should find it in the Public Library (this happened in St. Petersburg) and advised me to read the book very carefully.

I enquired at the Public Library, but the book (published in English) was not there. There was only a German translation of it in twenty parts, the first three being missing.

But soon I obtained the book elsewhere in English and actually found there certain hints very closely connected with my question, though they did not give a complete answer to it.

This instance, and a number of others like it, showed me that in these transitional states I went through the same experiences as do mediums, clairvoyants and the like. One voice told me something very interesting about the Temple of Solomon in Jerusalem, something that I thought I did not know before, or, if I had ever read it anywhere, had entirely forgotten. Among other things, in describing the temple, the voice said that there were *swarms of flies* there. Logically this was quite comprehensible and even inevitable. In a temple where sacrifices were made, where animals were killed and where there was certainly a great deal of blood and every kind of filth, there must undoubtedly have been many flies. At the same time this sounded new and, so far as I remember, I had never read of flies in connection with ancient temples. But not long before that I had been in the East myself and knew what quantities of flies can be there even under ordinary conditions.

These descriptions of Solomon's Temple, and particularly the " flies ", gave me a complete explanation of many strange things which I had come across in my reading and which I could call neither deliberate falsification nor real clairvoyance. Thus the " clairvoyance " of Leadbeater and Dr. Steiner, all the " Akashic records ", the descriptions of what happened tens of thousands of years ago in mythical Atlantis or in other prehistoric countries, were undoubtedly of the same nature as the flies in Solomon's Temple. The only difference was that I did not believe in my experiences, while the " Akashic records " were believed and are believed by both their authors and readers.

It very soon became evident to me that neither in these nor in the other experiences was there anything real. It was all reflected, it all came from the memory, from the imagination. The voices immediately became silent as soon as I passed to something familiar and concrete which could be verified.

This explained to me why it is that authors who describe Atlantis are unable with the aid of their " clairvoyance " to solve any practical problems relating to the present which are always so easy to find, but which for some reason they always avoid touching on. Why do they know everything that happened thirty thousand years ago and not know what is happening at the time of their experiments but in another place ?

During all these experiments I felt that if I were to believe these voices I should come to a standstill and go no further. This frightened me. I felt that it was all self-deception; that however inviting all that was said and promised by the voices might be, it would all lead nowhere, but would leave me exactly where I was. I understood that it was precisely this that was " beauty ", i.e. that it all came from the imagination.

I decided to struggle with these transitional states, adopting towards them a very critical attitude and rejecting as unworthy of credence *all that I might have imagined myself.* This immediately began to give results. As soon as I began rejecting everything I *heard,* realising it to be the same " stuff as dreams are made of ", and firmly discarded it for some time, refusing to listen to anything or pay attention to anything, my state and my experiences changed.

I passed the second threshold, which I have already mentioned, beyond which *a new world* began. The " voices " disappeared; in their place there sounded sometimes one voice, which could always be recognised whatever forms it might take. At the same time this new state differed from the transitional state by its extraordinary lucidity of consciousness. I then found myself in the world of mathematical relations, in which there was nothing at all resembling what occurs in life.

In this state also, after passing the second threshold and finding myself in the " world of mathematical relations ", I obtained answers to all my questions, but the answers often took a very strange form. In order to understand them it must be realised that the world of mathematical relations in which I was did not remain immovable ; this means—there was nothing in it that remained as it was the moment before. Everything moved, changed, was transformed and became something else. Sometimes I suddenly saw all mathematical relations disappear one after another into infinity. Infinity swallowed everything, filled everything; all distinctions were effaced. And I felt that one moment more and I myself should disappear into infinity. I was overcome with terror at the imminence of this abyss. Sometimes this terror made me jump to my feet, move about, in order to drive away the nightmare which had seized me. Then I felt that someone was laughing at me ; sometimes I seemed to hear the laugh. Suddenly I caught myself realising that it was I laughing at myself— that I had again fallen into the snare of " beauty ", that is, of a wrong approach. Infinity attracted me and at the same time frightened and repelled me. And I came to understand it quite differently. Infinity was not infinite continuation in one direction, but infinite variation at one point. I understood that the terror of infinity results from a wrong approach to it, from a wrong attitude to it. I understood that with a right approach to it infinity is precisely what explains everything, and that nothing can be explained without it.

At the same time I felt that in infinity there was a real menace and a real danger.

To describe consecutively the course of my experiences, the course of the ideas that came to me and the course of fleeting thoughts, is quite impossible, mainly because no one experiment was ever like another. Each time I learned something new about the same thing in such a way as fundamentally to alter all I had learned about it before.

A characteristic feature of the world in which I found myself was, as I have already said, its mathematical structure and the complete absence of anything that could be expressed in the language of ordinary concepts. To use the theosophical terminology I was in the *mental world* " *Arupa* ", but the peculiarity of my observations was that only this world " Arupa " really existed. All the rest was the creation of imagination. The real world was a " world without forms ". It is an interesting fact that in my first experiment I found myself probably at once or almost at once in this world, escaping the " world of illusions ". But in subsequent experiments " voices " seemed to try to detain me in the imaginary world, and I was able to get out of it only when I struggled firmly and resolutely with the illusions as they arose. All this strongly reminded me of something I had read before. It seemed to me that, in existing literature, in the descriptions of magical experiments or in the descriptions of initiations and preceding tests, there was something very similar to what I had experienced and felt—but of course this does not refer to modern " séances " or even to attempts at ceremonial magic, which is complete immersion in the world of illusions.

An interesting phenomenon in my experiments was the consciousness of danger which threatened me from infinity and the constant warnings received from *somebody*, as though there was *somebody* who watched me all the time and often tried to persuade me to stop my experiments, not to attempt to go along this path, which was wrong and unlawful from the point of view of certain principles which I at that time felt and understood only dimly.

What I have called "mathematical relations" were continually changing round me and within me, sometimes taking the form of sounds, of music, sometimes the form of a design, sometimes the form of light filling the whole of space, of a kind of visible vibration of light rays, crossing, interweaving with one another, pervading everything. In this connection there was an unmistakable feeling that through these sounds, through the design, through the light, I was learning something I had not known before. But to convey what I learned, to tell about it or put it into writing was very difficult. The difficulty of explaining was increased by the fact that words express badly, and really cannot express, the essence of the intense emotional state in which I was during these experiences.

This emotional state was perhaps the most vivid characteristic of the experiences which I am describing. Without it there would have been nothing. Everything came through it, that is, everything was understood through it. In order to understand my experiences it must be realised that I was not at all indifferent to the sounds and the light mentioned above. I took in everything through feeling, and experienced emotions which never exist in life. The new knowledge that came to me came when I was in an exceedingly intense emotional state. My attitude towards this new knowledge was in no way indifferent; I either loved it or was horrified by it, strove towards it or was amazed by it; and it was these very emotions, with a thousand others, which gave me the possibility of understanding the nature of the new world that I came to know.

The number "three" played a very important part in the world in which I found myself. In a way quite incomprehensible to our mathematics it entered into all the relations of magnitudes, created them and originated from them. All taken together, that is, the entire universe, sometimes appeared in the form of a "triad", composing one whole, and looking like some great trefoil. Each part of the "triad", by some inner process, was again transformed into a "triad", and this process continued until all was filled with "triads", which were transformed into music, or light, or designs. Once again I must say that all these descriptions express very badly what occurred, as they do not give the emotional element of joy, wonder, rapture, horror, continually changing one into the other.

As I have already said, the experiments were most successful when I was by myself and lying down. Sometimes, however, I tried being among people or walking in the streets. These experiments were usually unsuccessful. Something began, but ended almost at once, passing into a heavy physical state. But sometimes I found myself in another world. On such occasions the whole of the ordinary world changed in a very subtle and strange way. Everything became different, but it is absolutely impossible to describe what happened to it. The first thing that can be said is that there was nothing which remained indifferent for me. All taken together and each thing separately affected me in one way or another. In other words, I took everything emotionally, reacted to everything emotionally. Further, in this new world which surrounded me, there was nothing separate, nothing that had no connection with other things or with me personally. All things were connected with one another, and not accidentally, but by incomprehensible chains of causes and effects. All things were dependent on one another, all things lived in one another. Further, in this world there was nothing dead, nothing inanimate, nothing that did not think, nothing that did not feel, nothing unconscious. Everything was living, everything was conscious of itself. Everything spoke to me and I could speak to everything. Particularly interesting were the houses and other buildings which I passed, especially the old houses. They were living beings, full of thoughts, feelings, moods and memories. The people who lived in them were their *thoughts, feelings, moods*. I mean that the people in relation to the " houses " played approximately the same rôle which the different " I "s of our personality play in relation to us. They come and go, sometimes live in us for a long time, sometimes appear only for short moments.

I remember once being struck by an ordinary cab-horse in the Nevsky, by its head, its face. It expressed the whole being of the horse. Looking at the horse's face I understood all that could be understood about a horse. All the traits of horse-nature, all of which a horse is capable, all of which it is incapable, all that it can do, all that it cannot do ; all this was expressed in the lines and features of the horse's face. A dog once gave me a similar sensation. At the same time the horse and the dog were not simply horse and dog ; they were " atoms ", conscious, moving " atoms " of great beings— " the great horse " and " the great dog ". I understood then that we also are atoms of a great being, " the great man ". Each thing is an atom of a " great thing ". A glass is an atom of a " great glass ". A fork is an atom of a " great fork ".

This idea and several other thoughts that remained in my memory from my experiences entered into my book *Tertium Organum*, which was actually written during these experiments. Thus the formulations of the laws of the noumenal world and several other ideas referring to higher dimensions were taken from what I learned during these experiments.

Sometimes I felt during these experiments that I understood many things particularly clearly, and I felt that if I could in some way preserve what I understood at this moment, then I should know how to make myself pass into this state at any moment I might want it; I should know how to fix this state and how to make use of it.

The question as to how to fix this state arose continually and I put it to myself many times when I was in the state in which I could receive answers to my questions ; but I could never get a direct answer to it, that is, the answer which I wanted. Usually the answer began far away and, gradually widening, included everything, so that finally the answer to the question included the answers to all possible questions. Naturally, for that reason I could not retain it in my memory.

Once, I remember, in a particularly vividly-expressed new state, that is, when I understood very clearly all I wished to understand, I decided to find some formula, some key, which I should be able, so to speak, to throw across to myself for the next day. I decided to sum up shortly all I understood at that moment and write down, if possible in one sentence, what it was necessary to do in order to bring myself into the same state immediately, by one turn of thought without any preliminary preparation, since this appeared possible to me all the time. I found this formula and wrote it down with a pencil on a piece of paper.

On the following day I read the sentence, " Think in other categories ". These were the words, but what was their meaning ? Where was everything I had associated with these words when I wrote them ? It had all disappeared, had vanished like a dream. Certainly the sentence " think in other categories " had a meaning ; only I could not recollect it, could not reach it.

Later on exactly the same thing happened with this sentence as had happened with many other words and fragments of ideas that had remained in the memory after my experiences. In the beginning, these sentences seemed to be entirely empty. I even laughed at them, finding in them complete proof of the impossibility of transferring anything from there to here. But gradually something began to revive in my memory, and in the course of two or three weeks I remembered more and more of what was connected with these words. And though all of it still remained very vague, as if seen from afar, I began to see meaning, that is, special meaning, in words which in the beginning seemed merely abstract designations of something without any practical significance.

The same process was repeated almost every time. On the day

after the experiment I remembered very little. Sometimes towards evening some vague memories began to return. Next day I could remember more; during the following two or three weeks I was able to recollect separate details of the experiences, though I was always perfectly aware that in general only an infinitesimal part was remembered. When I tried to make experiments more often than every two or three weeks, I spoiled the results, that is, everything was confused, I could remember nothing.

But I will continue the description of successful experiments. Many times, perhaps always, I had the feeling that when I passed the second threshold I came into contact with *myself*, with the self which was always within me, which always saw me and always told me something that I could not understand and could not even hear in ordinary states of consciousness.

Why can I not understand?

I answered: merely because in the ordinary state thousands of voices sound at once creating what we call our "consciousness", our thoughts, our feelings, our moods, our imagination. These voices drown the sound of that inner voice. My experiments added nothing to the ordinary "consciousness"; they *reduced* it, yet by reducing it they intensified it to an incomprehensible degree. What did they actually do? They compelled these other voices of the ordinary consciousness to keep silence, put them to sleep, or made them inaudible. Then I began to hear the other voice, which came as it were from above, from a certain point *above my head*. I understood then that the whole problem and the whole object consisted in being able to hear this voice *constantly*, in being in constant communication with it. The being to whom this voice belonged knew everything, understood everything and above all was free from thousands of small and distracting "personal" thoughts and moods. He could take everything calmly, could take everything objectively, as it was in reality. And at the same time *this was I*. How this could be so and why in the ordinary state I was so far from myself, if this was I—that I could not explain. Sometimes during the experiments I called my ordinary self "I" and the other one—"he". Sometimes, on the contrary, I called the ordinary self "he" and the other one—"I". But I shall return later to the problem of "I" in general and the realisation of "I" in the new state of consciousness, because all this was much more complicated than the mere superseding of one "I" by the other.

At present I want to try to describe, so far as it has been preserved in my memory, how this "he" or this "I" looked at things as distinct from an ordinary "I".

I remember once sitting on a sofa smoking and looking at an ash-tray. It was an ordinary copper ash-tray. Suddenly I felt that I was beginning to understand what the ash-tray was, and at the same time, with a certain wonder and almost with fear, I felt that I had never understood it before and that we do not understand the simplest things around us.

The ash-tray roused a whirlwind of thoughts and images. It contained such an infinite number of facts, of events ; it was linked with such an immense number of things. First of all, with everything connected with smoking and tobacco. This at once roused thousands of images, pictures, memories. Then the ash-tray itself. How had it come into being ? All the materials of which it could have been made ? Copper, in this case—what was copper ? How had people discovered it for the first time ? How had they learned to make use of it ? How and where was the copper obtained from which this ash-tray was made ? Through what kind of treatment had it passed, how had it been transported from place to place, how many people had worked on it or in connection with it ? How had the copper been transformed into an ash-tray ? These and other questions about the history of the ash-tray up to the day when it had appeared on my table.

I remember writing a few words on a piece of paper in order to retain something of these thoughts on the following day. And next day I read :

" *A man can go mad from one ash-tray.*"

The meaning of all that I felt was that in one ash-tray it was possible to know *all*. By invisible threads the ash-tray was connected with everything in the world, not only with the present, but with all the past and with all the future. To know an ash-tray meant to know all.

My description does not in the least express the sensation as it actually was, because the first and principal impression was that the ash-tray was alive, that it thought, understood and told me all about itself. All I learned I learned from the ash-tray itself. The second impression was the extraordinary emotional character of all connected with what I had learned about the ash-tray.

" Everything is alive," I said to myself in the midst of these observations ; " there is nothing dead, it is only we who are dead. If we become alive for a moment, we shall feel that everything is alive, that all things live, think, feel and can speak to us."

The case of the ash-tray reminds me of another instance in which the answer to my question came in the form of a visual image, very characteristic in its structure.

Once when I was in the state into which my experiments brought me, I asked myself : " What is the world ? "

Immediately I saw a semblance of some big flower, like a rose or a lotus, the petals of which were continually unfolding from the middle, growing, increasing in size, reaching the outside of the flower and then in some way again returning to the middle and starting again at the beginning. Words in no way express it. In this flower there was an incredible quantity of light, movement, colour, music, emotion, agitation, knowledge, intelligence, mathematics, and continuous unceasing growth. And while I was looking at this flower *someone* seemed to explain to me that this was the " World " or " Brahma " in its clearest aspect and in the nearest possible approximation to what it is in reality—" If the approximation were made still nearer, it would be Brahma himself, as he is," said the voice.

These last words seemed to contain a kind of warning, as though Brahma in his real aspect was dangerous and could swallow up and annihilate me. This again was " infinity ".

This incident and the symbol of Brahma or " the world ", which remained in my memory, greatly interested me because it explained to me the origin of other symbols and allegorical images. I thought later that I understood the principle of the formation of the different attributes of gods and the meaning of many myths.

Moreover, this incident brings me to another very important feature of my experiments, namely, to the method by which ideas were communicated to me in these strange states after the second threshold.

As I have already said, ideas were transmitted not in words but in sounds, forms, " designs " or symbols. Usually everything began with the appearance of these forms. As was mentioned before, " voices " were the characteristic feature of the transitional state. When they ceased they were replaced by these forms, i.e. sounds, " designs ", etc. ; and after these followed visual images possessing very special properties and demanding detailed explanation. Brahma seen as a flower might serve as an example of these visual images, though ordinarily they were much simpler, something in the nature of conventional signs or hieroglyphs.

These signs constituted the form of speech or thought, or of what corresponded to speech or thought, in the state of consciousness which I attained. Signs or hieroglyphs moved and changed before me with dizzy rapidity, expressing in this way transitions, changes, combinations and correlations of ideas. Only this manner of " speech " was sufficiently quick for the quickness of thought which was here arrived at. No other forms were quick enough. And these *moving signs of things* indicated the beginning of new thinking, a new state of consciousness. Thinking in words became quite impossible. As I have already said, between two words of the same sentence long periods of time passed. Thinking in words could never keep pace with thought as it worked in this state.

It is curious that in mystical literature a number of references to these " signatures of things " can be found. I give them the name which was given to them by Jacob Boehme (*Tertium Organum*, Ch. XXII, p. 281). I do not doubt that Boehme spoke of exactly the same signs that I saw. For myself I call them " Symbols ". By their outer form it would be more correct to call them moving hieroglyphs. I tried to draw some of them and, though I sometimes succeeded in it, on the following day it was very difficult to connect the figures obtained with any ideas. Once, however, I obtained something very interesting.

I drew a figure like this :

FIG. 6.

The number of lateral projections is immaterial, but the important point is that they are disposed at unequal distances from one another along the horizontal line.

I obtained this figure in the following way.

In connection with certain facts in the lives of people whom I knew, which happened to come into my mind, I asked myself the rather complicated question as to how the fate of one man might influence the fate of another man. I cannot now reconstruct my question exactly, but I remember that it was connected with the idea of the laws of cause and effect, of free choice or accident. While still continuing to think in an ordinary way, I imagined the life of a man I knew and the accident in his life through which he had come across other people whose lives he had most decisively influenced, and who in their turn had changed many things in his own life. Thinking in this way, I suddenly noticed, or caught myself seeing, all these intercrossing lives in the form of simple signs, namely in the form of short lines with small projections on one side. The number of these projections diminished or increased ; they either approached one another or separated. And in their appearance, in their approach or separation, and also in the combination of different lines with different projections, were expressed the ideas and laws governing men's lives.

I will return later to the meaning of this symbol. At present I wish only to explain the actual method of obtaining new ideas in the state of consciousness described.

A separate part of my experiences constituted what I could call my relation to myself, or more correctly to my body. It all became alive, became thinking and conscious. I could speak to any part of my body as if it was a separate being, and could learn from it what attracted it, what it liked, what it disliked, what it was afraid of, what it lived by, what were its interests and needs. These conversations with the consciousnesses of the physical body revealed a whole new world.

I have tried to describe some of the results of these impressions in *Tertium Organum*, in speaking of consciousness *not parallel to our own*.

These consciousnesses, which I now call the consciousnesses of the physical body, had very little in common with our consciousness which objectivises the external world and distinguishes " I " from " not I " : These consciousnesses, i.e. the consciousnesses of the physical body, were completely immersed in themselves. They knew only themselves, only " I " ; " not I " did not exist for them. They could think only of themselves—they could speak only of themselves. But, as against that, they knew everything about themselves that could be known. I then understood that their nature and the form of their existence consisted in their continually speaking of themselves —what they were, what they needed, what they wished, what was pleasant for them, what was unpleasant, what dangers threatened them, what could ward off or remove these dangers.

In the ordinary state we do not hear these voices separately ; only the noise produced by them or their general tone is felt by us as our physical state or mood.

I have no doubt that if we could consciously enter into communication with these " beings " we should be able to learn from them all the details of the state of every function of our organism. The first idea that comes to one's mind in this connection is the consideration that this would be particularly useful, in the case of diseases and functional disorders, for right diagnosis, for the prevention of possible illnesses and for the treatment of those already existing. If a method could be found for entering into communication with these consciousnesses and for receiving from them information as to the state and demands of the organism, medicine would stand on firm ground.

In continuing my experiments I tried all the time to find a means of passing from abstract to concrete facts. I wanted to find out whether there was not a possibility of strengthening the ordinary powers of perception or of discovering new powers, especially with regard to events in time, to the past or future. I definitely put myself the question, whether the power can exist of seeing without the aid of eyes, or at a great distance, or through a wall, or of seeing things in closed receptacles, reading letters in sealed envelopes, reading a book on a shelf between other books, and so on. It had never been clear to me whether such things were possible. On the contrary, I knew that all attempts at verification of the phenomena of clair-voyance, which are sometimes described, invariably ended in failure.

During my experiments I many times attempted to " see ", for instance, when I was myself in the house, what was happening in the street, which I could not see in the natural way, or to " see " some man or other whom I knew well, what he was doing at that moment; or to reconstruct fully scenes from the past of which I knew only some parts.

Then I sealed some old photographs from an album into envelopes of the same size, mixed them up and tried to " see " whose portrait I held in my hand. I tried the same thing with playing-cards.

When I became convinced that I was not succeeding, I tried to reconstruct as a clear visual image what was undoubtedly in my memory, though in the ordinary state I could not visualise it at will. For instance I tried to " see " the Nevsky, beginning from Znamensky Square, with all the houses and shop-signs in their order. But this also was never successful when done intentionally. Unintentionally and in various circumstances I more than once saw myself walking along the Nevsky, and then I " saw " both the houses and the signs exactly as they would be in reality.

Finally I had to recognise as unsuccessful all attempts to pass to concrete facts. Either it is quite impossible, or else I attempted it in the wrong way.

But there were two cases which showed that there is a possibility of a very great strengthening of our capacities of perception in relation to the ordinary events of life.

Once I obtained not exactly clairvoyance, but undoubtedly a very great strengthening of the capacity of vision. It was in Moscow in the street, half an hour after an experiment which had seemed to me to be entirely unsuccessful. For a few seconds my vision suddenly became extraordinarily acute. I could quite clearly see the faces of people at a distance at which normally one would have difficulty in distinguishing one figure from another.

Another instance occurred during the second winter of my experiments in St. Petersburg. Circumstances were such that the whole of that winter I was unable to go to Moscow, although at the time I very much wanted to go there in connection with several different matters. Finally I remember that about the middle of February I definitely decided that I would go to Moscow for Easter. Soon after this I again began my experiments. Once, quite accidentally, when I was in the state in which moving signs or hieroglyphs were beginning to appear, I had a thought about Moscow, or about someone whom I had to see there at Easter. Suddenly without any warning I received the comment that I should not go to Moscow at Easter. Why? In answer to this I saw how, starting from the day of the experiment I have described, events began to develop in a definite order and sequence. Nothing new happened. But the causes, which I could see quite well and which were all there on the day of my experiment, were evolving, and having come to the results which unavoidably followed from them, they formed just before Easter a whole series of difficulties which in the end prevented me going to Moscow. The fact in itself, as I looked at it, had a merely curious character, but the interesting side of it was that I saw what looked like a possibility of calculating the future—the whole future was contained in the present. I saw that all that had happened before Easter resulted directly from what had already existed two months earlier.

Then in my experiment I probably passed on to other thoughts, and on the following day I remembered only the bare result, that " somebody " had told me I should not go to Moscow at Easter. This was ridiculous, because I saw nothing that could prevent it. Then I forgot all about my experiment. It came to my memory again only a week before Easter, when suddenly a whole succession of small circumstances brought it about that I did not go to Moscow. The circumstances were precisely those which I had " seen " during my experiment, and they quite definitely resulted from what had existed two months before that. Nothing new had happened.

When everything fell out exactly as I had seen, or foreseen, in that strange state, I remembered my experiment, remembered all the details, remembered that I saw and knew then what had to happen.

In this incident I undoubtedly came into contact with the possibility of a different vision in the world of things and events. But, speaking generally, all the questions which I asked myself referring to real life or to concrete knowledge led to nothing.

I think that this is connected with a principle which became clear to me during my experiments.

In ordinary life we think by thesis and antithesis; always and everywhere there is " yes " or " no ", " no " or " yes ". In thinking differently, in thinking in a new way, in thinking by means of signs of things, I came to understand the fundamental errors of our mental process.

In reality, everywhere and in every case there were not two but three elements. There were not only " yes " and " no ", but " yes ", " no " and something else besides. And it was precisely the nature of this " third " element, inaccessible to the understanding, which made all ordinary reasonings unsuitable and demanded a change in the basic method. I saw that the solution of all problems always came from a *third*, unknown, element, that is to say, it came from a third and unknown side, and that without this *third* element it was impossible to arrive at a right solution.

Further, when I asked a question I very often began to see that the question itself was wrongly put. Instead of giving an immediate answer to my question, the " consciousness " to which I was speaking began to move my question round and turn it about, showing me that it was wrong. Gradually I began to see what was wrong. As soon as I understood clearly what was wrong in my question, *I saw the answer*. But the answer always included a *third element* which I could not see before, because my question was always built upon *two* elements only, thesis and antithesis. I formulated this for myself in the following way : that the whole difficulty lay in the putting of the question. If we could put questions rightly, we should know the answers. A question rightly put contains the answer in itself. But the answer will be quite unlike what we expect, it will always be on another plane, on a plane not included in the ordinary question.

In several cases in which I attempted to think with certain ready-made words or with ready-made ideas I experienced a strange sensation like a physical shock. Before me complete emptiness opened out, because in the real world with which I had come into contact there was nothing corresponding to these words or ideas. The sensation was very curious—the sensation of unexpected emptiness where I had counted upon finding something, which, if not solid and definite, would be at least existent.

I have already said that I found nothing corresponding to the theosophical " astral bodies ", or " astral world ", nothing corresponding to " reincarnation ", nothing corresponding to the " future life " in the ordinary sense of the word, that is, to one or another form of existence of the souls of the dead. All this had no meaning, and not only did it not express any truth, but *it did not directly contradict truth*. When I tried to introduce into my thoughts the questions connected with these ideas, there were no replies to them ; words remained only words *and could not be expressed by any hieroglyphs*.

The same thing happened with many other ideas, for example with the idea of " evolution " as it is understood in " scientific " thinking. It did not fit in anywhere and did not mean anything at all. There was no place for it in the world of realities.

I realised that I felt which ideas were alive and which were dead ; dead ideas were not expressed in hieroglyphs, they remained words. I found an enormous number of such dead ideas in the general usage of thought. Besides the ideas already mentioned, all so-called " social theories " belonged to the dead ideas. They simply did not exist. There were words behind which lay no reality ; similarly the idea of " justice ", as it is ordinarily understood in the sense of " compensation " or " retribution ", was utterly dead. One thing could never compensate for another, one act of violence never destroy the results of another act of violence. At the same time the idea of justice in the sense of " desire for the general good " was also dead. And, speaking generally, there was some great misunderstanding in this idea. The idea assumed that a thing could exist by itself and be " unjust ", that is, contradict a certain law ; but in the real world *everything was one*, and there were no two things that could contradict each other. And therefore there was nothing that could be called justice or injustice. The only difference that existed was between dead and living things. But this distinction was exactly what we did not understand, and though we strove to express the same idea in our language we hardly succeeded in doing so.

All these are only examples. In fact almost all the usual ideas and concepts by which people live proved to be *non-existent*.

With great amazement I became convinced that only a very small number of ideas corresponds to real facts, that is, actually exists. We live in an entirely unreal, fictitious world, we argue about non-existent ideas, we pursue non-existent aims, invent everything, even ourselves.

But as opposed to dead ideas which did not exist *anywhere*, there were on the other hand living ideas incessantly recurring always and everywhere and constantly present in everything I thought, learned and understood at that time.

First there was the idea of the *triad*, or the trinity, which entered into everything. Then a very important place was occupied and much was explained by the idea of the four elements : *fire, water, air* and *earth*. This was a real idea, and during the experiments, in the new state of consciousness, I understood how it entered into everything and was connected with everything through the triad. But in the ordinary state the significance and connection of these two ideas eluded me.

Further, there was the idea of *cause and effect*. As I have already mentioned, this idea was expressed in hieroglyphs in a very definite way. But it was in no way connected with the idea of " reincarnation ", and referred entirely to ordinary earthly life.

A very great place—perhaps the chief place—in all that I had learned was occupied by the idea of " I ". That is to say, the feeling or sensation of " I " in some strange way changed within me. It is very difficult to express this in words. Ordinarily we do not sufficiently understand that at different moments of our life we feel our " I " differently. In this case, as in many others, I was helped by my earlier experiments and observations of dreams. I knew that in sleep " I " is felt differently, not as it is felt in a waking state ; just as differently, but in quite another way, " I " was felt in these experiences. The nearest possible approximation would be if I were to say that everything which is ordinarily felt as " I " became " not I ", and everything which is felt as " not I " became " I ". But this is far from being an exact statement of what I felt and learned. I think that an exact statement is impossible. It is necessary only to note that the new sensation of " I " during the first experiments, so far as I can remember it, was a very terrifying sensation. I felt that I was disappearing, vanishing, turning into nothing. This was the same terror of infinity of which I have already spoken, but it was reversed : in one case it was All that swallowed me up, in the other it was Nothing. But this made no difference, because All was equivalent to Nothing.

But it is remarkable that later, in subsequent experiments, the same sensation of the disappearance of " I " began to produce in me a feeling of extraordinary calmness and confidence, which nothing can equal in our ordinary sensations. I seemed to understand at that time that all the usual troubles, cares and anxieties are connected with the usual sensation of " I ", result from it, and, at the same time, constitute and sustain it. Therefore, when " I " disappeared, all troubles, cares and anxieties disappeared. When I felt that I did not exist, everything else became very simple and easy. At these moments I even regarded it as strange that we could take upon ourselves so terrible a responsibility as to bring " I " into everything and start from " I " in everything. In the idea of " I ", in the sensation of " I ", such as we ordinarily have, there was something almost abnormal, a kind of fantastic conceit which bordered on blasphemy, as if each one of us called himself God. I felt then that only God could call himself " I ", that only God was " I ". But we also call ourselves " I " and do not see and do not notice the irony of it. "

As I have already said, the strange experiences connected with

my experiments began with the change in the sensation of " I ",
and it is difficult to imagine that they would be possible in the case
of retention of the ordinary sensation of " I ". This change constituted their very essence, and everything else that I felt and learned
depended upon it.

With regard to what I learned during my experiments, particularly
with regard to the increase of the possibility of cognition, I came to
know much that was strange and that did not enter into any theories
that I had known before.

The consciousness which communicated with me by means of
moving hieroglyphs attached the greatest importance to this question
and strove to impress on my mind, perhaps more than anything else,
all that related to this question, that is, to the methods of cognition.

I mean that the hieroglyphs explained to me that besides the
ordinary cognition based on the evidence of the sense organs, on
calculation and on logical thinking, there exist *three other different
cognitions*, which differ from one another and from the ordinary cognition, not in degree, not in form, not in quality, but in their very
nature, as phenomena of utterly different orders which cannot be
measured by the same measure.   In our language we call these three
phenomena together, where we recognise their existence, intensified
cognition, that is, we admit their difference from the ordinary cognition,
but do not understand their difference from one another.   This,
according to the hieroglyphs, is the chief factor in preventing us from
understanding rightly our relation to the world.

Before attempting to define the " three kinds of cognition " I
must remark that the communication about the forms of cognition
always began from some question of mine which had no definite
relation to the problems of cognition, but evidently contradicted in
some way laws of cognition that were unknown to us.   For example,
this nearly always happened when from the domain of abstract questions I tried to pass to concrete phenomena, asking questions referring
to living people or real things, or to myself in the past, present or
future.

In those cases I received the answer that what I wished to know
could be known in three ways or that, speaking generally, there
were three ways of cognition, apart of course from the ordinary way
of cognition with the help of the sense-organs, calculation and logical
reasoning, which did not enter into the question, and the limits of
which were assumed to be known.

Further, there usually followed a description of the characteristics
and properties of each way.

It was as though someone anxious to give me right ideas of things found it particularly important that I should understand *this* rightly.

I will try to set forth as exactly as possible all that refers to this question. But I doubt whether I shall succeed in fully expressing even what I understand myself.

The first cognition is learning in an unusual way, as though through inner vision, anything relating to things and events with which I am directly connected and in which I am directly and personally interested. For instance, if I learn something which must happen in the near future to me or to someone closely connected with me, and if I learn it not in the ordinary way but through inner vision, this would be cognition of that kind. If I learn that a steamer on which I have to sail will be wrecked, or if I learn that on a definite day serious danger will threaten one of my friends, and if I learn that by taking such and such a step I can avert the danger—this will be cognition of the first kind or *the first cognition*. Personal interest constitutes a necessary condition of this cognition. Personal interest connects a man in a certain way with things and events and enables him to occupy in relation to them a definite " position of cognition ". Personal interest, that is, the presence of the person interested, is an almost necessary condition of " fortune-telling ", " clairvoyance ", " prediction of the future " ; without personal interest these are almost impossible.

The second cognition is also cognition of ordinary things and events in our life, for knowing which we have no ordinary means—just as in the first case—but with which nothing connects us personally. If I learn that a steamer will be wrecked, in the fate of which I am not personally interested at all, on which neither I nor any of my friends is sailing ; if I learn that which is happening in my neighbour's house, but which has no relation to myself; if I learn for certain who actually were the persons who are considered historical enigmas, like the Man in the Iron Mask or Dmitry the Pretender or the Comte de Saint-Germain, or if I learn somebody's future or past, again having no relation to myself, this will be the second kind of cognition. The second kind of cognition is the most difficult, and is almost impossible, because if a man accidentally, or with the aid of special means or methods, learned more than other people can know he would certainly do so in the first way.

The second kind of cognition contains something unlawful. It is " magic ", in the full sense of the word. The first and third ways of cognition in comparison with it appear simple and natural, though the first way, based on emotional apprehension, presentiment or desire of some kind, looks like a psychological trick ; and the third way appears as a continuation of ordinary cognition, but along new lines and on new principles.

The third cognition is cognition based on knowledge of the mechanism of everything existing. By knowing all the mechanism and by knowing all the relations of the separate parts, it is easy to arrive at the smallest detail and determine with absolute precision everything connected with this detail. The third cognition is cognition based on calculation. Everything can be calculated. If the mechanism of everything is known it is possible to calculate what kind of weather there will be in a month's time, or in a year's time; it would be possible to calculate the day and hour of every occurrence. It would be possible to calculate the meaning and significance of every small event that is observed. The difficulty of the third order of cognition consists first in the necessity of knowing the whole mechanism for the cognition of the smallest thing, and second, in the necessity for putting into motion the whole colossal machine of knowledge in order to know something quite small and insignificant.

This is roughly what I " learned " or " understood " in reference to the three kinds of cognition. I see quite clearly that in this description the idea is inadequately conveyed; many things, probably the most important, escaped my memory long ago. This is true not only in relation to the question of cognition, but, generally, in relation to all that I have written here about my experiments. All these descriptions must be taken very cautiously, on the understanding that in the description, ninety-nine per cent. of what was felt and understood during the experiments has been lost.

A very strange place in my experiments was occupied by attempts to know something concerning the dead. Questions of this kind usually remained without an answer, and I was vaguely aware that there was some essential fault in the questions themselves. But once I received a very clear answer to my question. Moreover, this answer was associated with another case of unusual sensation of death, which I experienced about ten years before the experiments described and which was caused by a state of intense emotion.

In speaking of both cases I shall have to touch on entirely personal matters.

The experience was connected with the death of a certain person closely related to me. I was very young at the time and was very much depressed by his death. I could not think of anything else and was trying to understand, to solve the riddle of disappearance and of men's interconnection with one another. And suddenly within me there rose a wave of new thoughts and new sensations, leaving after it a feeling of astonishing calm. I saw for a moment why we cannot understand death, why death frightens us, why we cannot find answers to any questions which we put to ourselves in connection with the problem of death. This person who had died, and of whom I was thinking, could not have died because he had never existed. This was the solution. Ordinarily, I had seen not him himself, but something that was like his shadow. The shadow had disappeared. The man who had really existed could not have disappeared. He was bigger than I had seen him, "longer", as I formulated it to myself, and in this "length" of his there was contained, in a certain way, the answer to all the questions.

This sudden and vivid current of thought disappeared as quickly as it had appeared. For a few seconds only there remained of it something like a mental picture. I saw before me two figures. One, quite small, was like the vague silhouette of a man. This figure represented the man as I had known him. The other figure was like a road in the mountains which you see winding among the hills, crossing rivers and disappearing into the distance. This was what he had been in reality and this was what I could neither understand nor express. The memory of this experience gave me for a long time a feeling of calm and confidence. Later, the ideas of higher dimensions gave me the possibility of finding a formulation for this strange "dream in a waking state", as I called my experience.

Something closely resembling this happened again in connection with my experiments.

I was thinking about another person also closely related to me who had died two years before. In the circumstances of this person's death, as also in the events of the last years of his life, there was much that was not clear to me, and there were things for which I might have blamed myself psychologically, chiefly for my having drifted away from him, not having been sufficiently near him when he might have needed me. There was much to be said against these thoughts, but I could not get rid of them entirely, and they again brought me to the problem of death and to the problem of the possibility of a life beyond the grave.

I remember saying to myself once during the experiment that if I believed in "spiritualistic" theories and in the possibility of communication with the dead I should like to see this person and ask him one question, just one question.

And suddenly, without any preparation, my wish was satisfied, and I *saw* him. It was not a visual sensation, and what I saw was not his external appearance, but *the whole of his life*, which flashed quickly before me. This life—this was he. The man whom I had known and who had died had never existed. That which existed was something quite different, because his life was not simply a series of events, as we ordinarily picture the life of a man to ourselves, but a thinking and feeling *being* who did not change by the fact of his death. The man whom I had known was the *face*, as it were, of this being—the face which changed with the years, but behind which stood always the same unchanging reality. To express myself figuratively I may say that I saw the man and spoke to him. In actual fact there were no visual impressions which could be described, nor anything like ordinary conversation. Nevertheless, I know that it was *he*, and that it was he who communicated to me much more about himself than I could have asked. I saw quite clearly that the events of the last years of his life were as inseparably linked with him as the features of his face which I had known during his life. These events of the last years were the features of the face of his life of the last years. Nobody could have changed anything in them, just as nobody could have changed the colour of his hair or eyes, or the shape of his nose ; and just in the same way it could not have been anybody's fault that this man had these facial features and not others.

The features of his face, like the features of his life of the last years—these were his qualities, these were he. To regard him without the events of the last years of his life would have been just as strange as to imagine him with a different face—it would not have been he. At the same time I understood that nobody could be responsible that he was as he was and not different. I realised that we depend upon one another much less than we think. We are no more responsible for the events in one another's lives than we are for the features of one another's faces. Each has his own face, with its own peculiar lines and features, and each has his own fate, in which another man may occupy a certain place, but in which he can change nothing.

But having realised this I saw also that we are far more closely bound to our past and to the people we come into contact with than we ordinarily think, and I understood quite clearly that death does not change anything in this. We remain bound with all with whom we have been bound. But for communication with them it is necessary to be in a special state.

I could explain in the following way the ideas which I understood in this connection : if one takes the branch of a tree with the twigs, the cross-section of the branch will correspond to a man as we ordinarily see him ; the branch itself will be the life of the man, and the twigs will be the lives of the people with whom he comes into contact.

The hieroglyph described earlier, a line with lateral projections, signifies precisely this branch with twigs.

I have endeavoured in my book *Tertium Organum* to set forth the idea of the "long body" of man from birth to death. The term used in Indian philosophy, "Linga Sharira", designates precisely this "*long body of life*".

The conception of man or the life of man as a branch, with offshoots representing the lives of people with whom he is connected, linked together many things in my understanding and explained a great deal to me. Each man is for himself such a branch, other people with whom he is connected are his offshoots. But each of these people is for himself a main branch and the first man for him is his offshoot. Each of the offshoots, if attention is concentrated upon it, becomes itself a branch with offshoots. In this way the life of each man is connected with a number of other lives, one life enters, in a sense, into another, and all taken together forms a single whole, the nature of which we do not understand.

This idea of the unity of everything, in whatever sense and on whatever scale it be taken, occupied a very important place in the conception of the world and of life that was formed in me in these strange states of consciousness. This conception of the world included something entirely opposed to our ordinary view of the world or conception of the world. Ordinarily each thing and each event has for us some value of its own, some significance of its own, some meaning of its own. This separate meaning that each thing, each event, has, is much more comprehensible and familiar to us than its possible general meaning and general significance, even in cases in which we can suppose or think of this general significance. But in this new conception of the world everything was different. Each thing appeared, first of all, not as a separate whole, but as a part of another whole, in most cases incomprehensible and unknown to us. The meaning and significance of the thing were determined by the nature of this great whole and by the place which it occupied in this whole. This completely changed the entire picture of the world. We are accustomed to take everything separately. Here there was nothing separate, and it was extraordinarily strange to feel oneself in a world in which all things were connected one with another and all things followed one from another. Nothing existed separately. I felt that the separate existence of anything—including myself—was a fiction, something non-existent, impossible. The sensation of absence of separateness and the sensation of connectedness and oneness united with the emotional part of my conceptions. At the beginning the combined sensation was felt as something terrifying, oppressive and hopeless ; but later, without changing its nature, it began to be felt as the most joyous and radiant sensation that could exist.

Further, there was a picture or mental image which entered into everything and appeared as a necessary part of every logical or illogical construction. This image showed two aspects, both of everything taken together, that is, the whole world, and of every separate part of it, that is, each separate side of the world and of life. One aspect was connected with the First Principle. I saw, as it were, the origin of the whole world or the origin of any given phenomenon or any given idea. The other aspect was connected with separate things : I saw the world, or those events which interested me at the particular moment, in their final manifestation, that is, as we see them around us, but connected into a whole, incomprehensible to us. But between the first aspect and the second aspect there always occurred an interruption like a gap or blank space. Graphically I might represent this approximately in the following way : Imagine that from above three lines appear from one point ; each of these three lines is again transformed into three lines ; each of these three lines again into three lines. Gradually the lines break more and more and gradually become more and more varied in properties, acquiring colour, form and other qualities, but not reaching real facts, and transforming themselves into a kind of invisible current proceeding from above. From below, imagine the infinite variety of phenomena collected and classified into groups ; these groups again unite, and as a result great numbers of very varied phenomena are actually bound into wholes and can be expressed by one sign or one hieroglyph. A series of these hieroglyphs represents life or the visible world at a certain distance from the surface. From above goes the process of differentiation, and from below goes the process of integration. But differentiation and integration do not meet. Between what is above and what is below is formed a blank space in which nothing is visible. The upper differentiating lines, multiplying and acquiring different colours, merge quickly together and disappear into a blank space which separates what is above from what is below. From below all the infinitely varied phenomena are very soon transformed into principles, extraordinarily rich in meaning and in hieroglyphic designation, but nevertheless smaller than the last of the visible upper lines.

It was approximately in this graphic representation that these two aspects of the world and things appeared to me. Or I might say that both above and below the world was represented on different scales, and these scales never met for me, never passed into one another, remained entirely incommensurable. The whole difficulty was precisely in this, and this difficulty was felt all the time. I realised that if I could throw a bridge from what was below to what was above or, still better, in the opposite direction, from what was above to what was below, I should understand everything that was below, because starting from above, the fundamental principles, it would have been easy and simple to understand anything below. But I never succeeded in connecting principles with facts because, though, as I have already said, all the facts very quickly became merged into complicated hieroglyphs, these hieroglyphs still differed very much from the *upper* principles.

Nothing that I am writing, nothing that can be said, about my experiences, will be comprehensible if the continuous emotional tone of these experiences is not taken into consideration. There were no calm, dispassionate, unexciting moments at all; everything was full of emotion, feeling, almost passion.

The strangest thing in all these experiences was the coming back, the return to the ordinary state, to the state which we call life. This was something very similar to dying or to what I thought dying must be.

Usually this coming back occurred when I woke up in the morning after an interesting experiment the night before. The experiments almost always ended in sleep. During this sleep I evidently passed into the usual state and awoke in the ordinary world, in the world in which we awake every morning. But this world contained something extraordinarily oppressive, it was incredibly empty, colourless and lifeless. It was as though everything in it was wooden, as if it was an enormous wooden machine with creaking wooden wheels, wooden thoughts, wooden moods, wooden sensations; everything was terribly slow, scarcely moved, or moved with a melancholy wooden creaking. Everything was dead, soulless, feelingless.

They were terrible, these moments of awakening in an unreal world after a real one, in a dead world after a living, in a limited world, cut into small pieces, after an infinite and entire world.

I did not obtain particularly new facts through my experiments, but I got many thoughts. When I saw that my first aim, i.e. objective magic, remained unattainable, I began to think that the artificial creation of mystical states might become the beginning of a new method in psychology. This aim would have been attained if I had found it possible to change my state of consciousness while at the same time retaining full power of observation. This proved to be impossible to the full extent. The state of consciousness changed, but I could not control the change, could never say *for certain* in what the experiment would result, and even could not always observe ; ideas followed upon one other and vanished too quickly. I had to recognise that though my experiments had established many possibilities, they did not give material for exact conclusions. The fundamental questions as to the relation of subjective magic to objective magic and to mysticism remained without decisive answers.

But after my experiments I began to understand many things differently. I began to understand that many philosophical and metaphysical speculations, entirely different in theme, form and terminology, might in actual fact have been attempts to express precisely that which I came to know, and which I have tried to describe. I understood that behind many of the systems of the study of the world and man there might lie experiences and sensations very similar to my own, perhaps identical with them. I understood that for centuries and thousands of years human thought has been circling and circling round something that it has never succeeded in expressing.

In any case my experiments established for me with indisputable clearness the possibility of coming into contact with the *real* world that lies behind the wavering mirage of the visible world. I saw that knowledge of the real world was possible but, as became clearer and clearer to me during my experiments, it required a different approach and a different preparation.

Putting together all that I had read and heard of, I could not but see that many before me had come to the same result, and many, most probably, had gone much further than I. But all of them had always been inevitably confronted with the same difficulty, namely the impossibility of conveying in the language of the dead the impressions of the living world. All of them except those who knew another approach. . . . I came to the conclusion that without the help of those who know another approach it is impossible to do anything.

1912–1929.

CHAPTER IV
ON THE STUDY OF DREAMS AND
ON HYPNOTISM

The strange life of dreams—"Psychoanalysis"—Impossibility of observing by usual methods—"Half-dream states"—Recurring dreams—Their simple nature—Dreams of flying—Dreams with staircases—False observations—Different degrees of sleep—Head dreams—Impossibility of pronouncing one's name in sleep—Different categories of dreams—Impersonation—Imitative dreams—Maury's dream—Development of dreams from end to beginning—Emotional dreams—Dream of Lermontoff—The building up of visual images—One man in two aspects—Material of dreams—The principle of " compensation "—The principle of complementary tones—Possibility of observing dreams in waking state—The sensation that " this has happened before."
Hypnotism—Hypnotism as means of bringing about the state of maximum suggestibility—The control of ordinary consciousness and logic, and impossibility of their complete disappearance—Phenomena of " mediumism "—Application of hypnosis in medicine—Mass hypnosis—The " rope trick "—Self-hypnosis—Suggestion—Necessity for studying these two phenomena separately—Suggestibility and suggestion—How duality is created in man—Two kinds of self-suggestion—Impossibility of voluntary self-suggestion.

POSSIBLY the most interesting first impressions of my life came from the world of dreams. And from my earliest years the world of dreams attracted me, made me search for explanations of its incomprehensible phenomena and try to determine the inter-relation of the real and the unreal in dreams. Certain quite extra-ordinary experiences were, for me, connected with dreams. When still a child I woke on several occasions with the distinct feeling of having experienced something so interesting and enthralling that all that I had known before, all I had come into contact with or seen in life, appeared to me afterwards to be unworthy of attention and devoid of any interest. Moreover, I was always struck by recurring dreams, dreams which occurred in the same form, in the same surroundings, led to the same results, to the same end, and always left behind the same feelings.

About 1900, when I had already read almost all I could find on dreams in psychological literature,[1] I decided to try to observe my

---

[1] In speaking of the literature on dreams I do not have in mind so-called psychoanalysis, that is, the theories of Freud and his followers, Jung, Adler and others. The reason for this is first, that when I began to be interested in dreams psychoanalysis was not yet in existence, or was very little known, and secondly, that, as I subsequently became convinced, there is and there was in psychoanalysis nothing of value, nothing that would make me alter the least of my conclusions though they are invariably all opposed to the psycho-analytical.

In order not to return again to this question I want to remark here that other aspects of psychoanalysis besides the unsuccessful attempt to study dreams are just as weak and dreams systematically.

My observations pursued a double purpose :

1. I wanted to collect as much material as possible for judging the structure and origin of dreams and I began, as is usually recommended, to write down my dreams immediately on awakening.

2. I wanted to verify a rather fantastic idea of my own which had made its appearance almost in my childhood : *was it not possible to preserve consciousness in dreams*, that is, to know while dreaming that one is asleep and *to think consciously* as we think when awake.

The first, that is, writing down dreams and so on, very soon often harmful, because they promise very much and there are people who believe in these promises and owing to this they completely lose the ability of distinguishing between the real and the false.

The only service psychoanalysis has rendered psychology as a whole is a precise formulation of the principle of the necessity of more and more observations in regions which so far have not entered into the subject of psychology. But it is exactly this principle which psychoanalysis itself has failed to follow because, having brought forward in the first stages of its existence a series of very doubtful hypotheses and generalisations, in the next stage it dogmatised them and in this way stopped any possibility of its own development. The specific " psychoanalytical " terminology which has grown out of these dogmatised hypotheses and become a kind of jargon helps us to recognise the adherents of psychoanalysis and their followers no matter how they call themselves and no matter how much they try to deny the connection between different schools and divisions of psychoanalysis and their origin in a common source.

The characteristic feature of this jargon is that it consists of words relating to non-existing phenomena which are accepted by the followers of psychoanalysis as existing. On the imaginary existence of these phenomena and on their imaginary relations to one another psychoanalysis has constructed a fairly complicated system something like the " natural philosophy " of the beginning of the 19th century, or like certain mediæval systems which also consisted in the description and classification of non-existing phenomena, as, for instance, various very exact and detailed *demonologies*.

The funny side of psychoanalysis, as a study of its history shows, is that all the principal features of the latest psychoanalysis were deduced by Dr. Freud on the basis of observations on *one case* in the middle eighties of last century. These observations of *one female patient* form the entire basis of psychoanalysis and of all its theories and, what is particularly interesting, these observations were made while using a method which was later condemned by Freud himself. The method consisted in hypnotising the patient and putting questions to her about herself which she could not answer in a normal state. As it has been established with an undoubted accuracy, both before and after this experiment, this method can lead to nothing because by persisting in questions of this kind either the hypnotiser without knowing it suggests answers to the hypnotised subject or the hypnotised subject invents fantastic theories and tells imaginary tales. In such a manner the famous " father complex " was found which brought along with it the " mother complex " and later on the whole box of tricks, the " Œdipus myth," etcetera.

The principal facts referring to this tragi-comic aspect of psychoanalysis can be found in a book by Stefan Zweig, one of the chief apologists of Freud. Fortunately the author brings out these facts obviously entirely without realising their significance.

The later tendency of psychoanalysis is to call itself *psychology* and to speak in the name of psychology in general.

The amusing side to this is that, under the mask of psychology psychoanalysis has penetrated into the domains of university science in several countries and forms a part of the compulsory curricula in some medical schools and faculties, so that students are obliged to undergo examinations in all this muddle.

The undoubted success of psychoanalysis in modern thought is explained by the poverty of the ideas, the timidity of the methods and the complete absence of inclination towards any practical application of its theories on the part of psychology which remains scientific, and then, most of all, by the very painfully felt need of a *general system*.

The popularity of psychoanalysis in certain literary and art circles and among certain classes of the public is explained by the justification and defence of homo-sexuality by psychoanalysis.

brought me to the understanding of the impossibility of a practical realisation of the usually recommended methods of observing dreams. *Dreams do not stand observation* ; *observations change them.* And I very soon noticed that I was observing, not those dreams which I used to have before, but new dreams *which were created by the very fact of observation.* There was something in me which at once began *to invent dreams* directly it felt that they were attracting attention. This made the usual methods of observation obviously useless.

The second, that is, *attempts to preserve consciousness in sleep*, created, most unexpectedly for me, a new way of observing dreams which I had not before suspected. Namely, they created a particular half-dream state. And I was very quickly convinced that without the help of half-dream states it was quite impossible to observe dreams without changing them.

" Half-dream states " began to appear probably as a result of my efforts to observe dreams at moments of falling asleep or in half-sleep after awaking. I cannot say exactly when these states began to come in full form. Probably they developed gradually. I think they began to appear for a short time before the moment of falling asleep, but if I allowed my attention to dwell on them I could not sleep afterwards. I came therefore gradually, by experience, to the conclusion that it was much easier to observe " half-dream states " in the morning, when already awake but still in bed.

Wishing to create these states, after waking I again closed my eyes and began to doze, at the same time keeping my mind on some definite image, or some thought. And sometimes in such cases there began those strange states which I call " half-dream states ". Without definite efforts such states would not come. Like all other people I either slept or did not sleep, but in these " half-dream states " I both slept and did not sleep at the same time.

If I take the time when these " half-dream states " were just beginning, i.e. when they came at the moment of going to sleep, then usually the first sign of their approach was the " hypnagogic hallucinations " many times described in psychological literature. I will not dwell on this. But when " half-dream states " began to occur chiefly in the morning, they usually started without being preceded by any visual impressions.

In order to describe these " half-dream states " and all that was connected with them, it is necessary to say a great deal. But I shall try to be as brief as possible because at the present moment I am concerned not with them but with their results.

The first sensation they produced was one of astonishment. I expected to find one thing and found another. The next was a feeling of extraordinary joy which the " half-dream states ", and the possibility of seeing and understanding things in quite a new way, gave me. And the third was a certain fear of them, because I very soon noticed that if I let them take their own course they would begin to grow and expand and encroach both upon sleep and upon the waking state.

Thus " half-dream states " attracted me on the one hand and frightened me on the other. I felt in them enormous possibilities and also a great danger. But what I became absolutely convinced of was that *without these " half-dream states " no study of dreams is possible* and that all attempts at such study are inevitably doomed to failure, to wrong deductions, to fantastic hypotheses, and the like.

From the point of view, therefore, of my original idea of the study of dreams I could be very content with the results obtained. I possessed a key to the world of dreams, and all that was vague and incomprehensible in them gradually cleared up and became comprehensible and visible.

The fact is that in " half-dream states " I was having all the dreams I usually had. But I was fully conscious, I could see and understand how these dreams were created, what they were built from, what was their cause, and in general what was cause and what was effect. Further, I saw that in " half-dream states " I had a certain control over dreams. I could create them and could see what I wanted to see, although this was not always successful and must not be understood too literally. Usually I only gave the first impetus, and after that the dreams developed as it were of their own accord, sometimes greatly astonishing me by the unexpected and strange turns which they took.

I had in " half-dream states " all the dreams I was able to have in the ordinary way. Gradually my whole repertoire of dreams passed before me. And I was able to observe these dreams quite consciously, could see how they were created, how they passed one into another, and could understand all their mechanism.

The dreams, observed in this way, became gradually classified and divided into definite categories.

To one of these categories I assigned all the constantly recurring dreams which I had had from time to time during the whole of my life from early childhood.

Some of these dreams used previously to frighten me by their persistence, their frequent repetition and a certain strange character, and made me look for a hidden or allegorical meaning, prophecy or warning in them. It had seemed to me that these dreams must have a certain significance, that they must refer to something in my life.

Speaking generally, naïve thinking about dreams always begins with the idea that all, and especially persistently recurring, dreams must have a certain meaning, must foretell the future, show the hidden traits of one's character, express physical qualities, inclinations, hidden pathological states, and so on.   In reality, however, as I very soon became convinced, my recurring dreams were in no way connected with any traits or qualities of my nature, or with any events in my life.   And I found for them clear and simple explanations which left no doubt as to their real nature.

I will describe several of these dreams with their explanations.

The first and most characteristic dream, which I had very often, was one in which I saw a quagmire or bog of a peculiar character which I was never able to describe to myself afterwards.   Often this quagmire or bog, or merely deep mud, such as is seen on Russian roads and even in Moscow streets, appeared before me on the ground or even on the floor of the room, without any association with the plot of the dream.   I did my utmost to avoid this mud, not to step into it, even not to touch it.   But I invariably got into it, and it began to suck me in and generally sucked my legs in up to the knees.   I made every conceivable effort to get out of this mud or mire, and sometimes I succeeded, but then I usually awoke.

It was very tempting to interpret this dream allegorically, as a threat or a warning.   But when I began to have this dream in " half-dream states " it was explained very simply.   The whole content of this dream was created by the sensation of my legs being entangled in the blanket or sheets, so that I could neither move nor turn them. If I succeeded in turning over, I escaped from this mud, but then I invariably woke up, because I made a violent movement.   As regards the mud itself and its " peculiar " character, this was connected, as I again became convinced in " half-dream states ", with the more imaginary than real " fear of bogs " I had in childhood.   This fear, which children and sometimes even grown-up people often have in Russia, is created by tales of quagmires and bogs and " windows ".[1] And in my case, observing this dream in " half-dream state " I could reconstruct where the sensation of the peculiar mud came from. This sensation and the visual images were quite definitely associated with tales of quagmires and " windows " which were said to have a " peculiar " character, that they could be recognised, that they always differed from an ordinary swamp, that they " sucked in " what fell into them, that they were filled with a *particular* soft mire, and so on, and so on.

---

[1] " Window " is the name given to a small spot, sometimes only a few yards across, of " bottomless " quagmire in an ordinary swamp.

In " half-dream states " the sequence of associations in the whole dream was quite clear. First appeared the sensation of bound legs, then the signal : bog, mire, window, *peculiar* soft mud. Then fear, desire to tear oneself away and usually the awakening. There was nothing, absolutely nothing, mystical or psychologically significant in these dreams.

Second, there was a dream which also frightened me. *I dreamed that I was blind.* Something was happening around me, I heard voices, sounds, noises, movement, felt some danger threatening me ; and I had to move somewhere with hands stretched out in front of me in order to avoid knocking against something, making all the time terrible efforts to see what was around me.

In " half-dream states " I understood that the effort I was making was not an effort to see, but an effort to open my eyes. And it was this effort, together with the *sensation of closed eyelids* which I could not lift, that created the sensation of " blindness ". Sometimes as the result of this effort I woke up. This happened when I actually succeeded in opening my eyes.

Even these first observations of recurring dreams showed me that dreams depend much more on the direct sensations of a given moment than on any general causes. Gradually I became convinced that almost all recurring dreams were connected with the sensation not even of a state, but simply with the sensation of the posture of the body at the given moment.

When I happened to press my hand with my knee and the hand became numb, I dreamt that a dog was biting my hand. When I wanted to take something in my hands or lift it, it fell out of my hands because my hands were as limp as rags and refused to obey me. I remember once in a dream I had to break something with a hammer, and the hammer was as if made of indiarubber ; it rebounded from the object I was striking, and I could not give any force to my blows. This, of course, was simply the sensation of relaxed muscles.

There was another recurring dream which always frightened me. In this dream I was a paralytic or a cripple ; I fell down and could not get up, because my legs did not obey me. This dream also seemed to be a presentiment of what was going to happen to me, until in " half-dream states " I became convinced that it was merely the sensation of motionless legs with relaxed muscles, which of course could not obey moving impulses.

Altogether I saw that our movements, especially our impulses to movements, and the sense of impotence in making a particular movement, play the most important rôle in the creation of dreams.

To the category of constantly recurring dreams belonged also dreams of flying. I used to fly fairly often and was very fond of these dreams. In "half-dream states" I saw that flying depended on a slight giddiness which occurs in sleep from time to time without any pathological cause, but probably simply in connection with the horizontal position of the body. There was no erotic element in the dreams of flying.

Amusing dreams which occur very often, those in which one sees oneself undressed or half-dressed walking in the street or among people, also required no complicated theories for their explanation. This was simply the sensation of one's half-dressed body. As I noticed in "half-dream states", these dreams occurred chiefly when I was feeling cold during sleep. The cold made me realise that I was undressed, and this sensation penetrated into my dreams.

Some of the recurring dreams could be explained only in connection with others. Such were the dreams of stairs, often described in psychological literature. These are strange dreams, and many people have them. You go up huge, gloomy, endless staircases, find certain passages leading out, remember the way, then lose it again, come upon unfamiliar landings, turnings, doors, etc. This is one of the most typical recurring dreams. And as a rule you meet no one, you are usually alone amidst these large empty staircases.

As I understood in "half-dream states", these dreams are a combination of two motives or recollections. The first motive is created by motor memory, the memory of direction. These dreams of stairs are in no way different from dreams of long corridors, with endless court-yards through which you pass, with streets, alleys, gardens, parks, fields, woods; in a word these are dreams of *roads* or *ways*. We all know many roads and ways; in houses, up stairs and along corridors; in towns, in the country, in the mountains; and we can see all these roads in dreams, although very often we see not the roads themselves but, if it can be so expressed, the general feeling of them. Each way has its own particular sensation. These sensations are created by thousands of small details reflected and impressed in various corners of our memory. Later these sensations are reproduced in dreams, though for the creation of the desired sensations dreams very often use the accidental material of images. Because of this the "road" you see in dreams may not resemble outwardly the road you actually know and remember when awake, but it will produce the same impressions as the road you know and are familiar with, and will give you the same sensations.

"Stairs" are similar "roads", only, as has already been said, they contain another motive as well. This motive consists in a certain mystical significance which stairs have in the life of every man. Everybody in his life often experiences on the stairs a sense of something new and unknown awaiting him that very moment on the next floor, behind a closed door. Everyone can recollect many such moments in his life. A man ascends the stairs not knowing what awaits him. For children it is often their arrival at school, or generally the first impression of school, and such impressions remain throughout life. Further, stairs are often the scene of hesitations, decisions, change of decisions, and so on. All this taken together and united with memories of motion creates dreams about staircases.

To continue the general description of dreams, I must note that visual images in sleep often do not correspond to visual images in waking states. A man you know very well in life can look quite different in a dream. In spite of that, however, you do not doubt for a minute that it is really he, and his unfamiliar aspect does not surprise you in the least. It often happens that the quite fantastic, and even unnatural and impossible, aspect of a man expresses certain traits and qualities you know in him. In a word, the outward form of things, people and events in dreams is much more plastic than it is in a waking state and it is much more susceptible to the influence of the accidental thoughts, feelings and moods that pass through us.

As regards recurring dreams, their simple nature and the absence of any allegorical meaning in them became quite unquestionable for me after they had all occurred several times in my " half-dream states ". I saw how they began, I could explain clearly where they came from and how they were created.

There was only one dream which I was unable to explain. That was the dream in which I saw myself *running on all fours*, and sometimes very fast. It seemed to be in certain cases the swiftest, safest and most reliable means of locomotion. In a moment of danger, or in general in any difficult situation, I always preferred in the dream this means of locomotion to any other.

For some reason I do not remember this dream in " half-dream states ". And I understood the origin of this " running on all fours " only later when I was observing a small child who was only just beginning to walk. He could walk, but to him it was still a great adventure and his position on two legs was still very uncertain, unstable and unreliable. He apparently distrusted himself in this position. If therefore anything unexpected happened, if a door opened, or a noise was heard from the street, or even if the cat jumped off the sofa, he dropped immediately on all fours. In observing him I understood that somewhere, deep in the innermost recesses of our memory, are preserved recollections of these first motor impressions and of all the sensations, fears and motor impulses connected with them. Evidently there was a time when new and unexpected impressions created the impulse to drop on all fours, that is, to assume a steadier and firmer position. In a waking state this impulse is not sufficiently strong, but it acts in dreams and creates strange pictures, which had also appeared to me to be allegorical or to have some hidden meaning.

Observations of the same child also explained to me a great deal about staircases. When he began to feel quite sure of himself on the floor, the stairs were still for him a great adventure. And nothing attracted him more than the stairs. Besides, he was forbidden to go near them. And of course in the next period of his life he practically lived on the stairs. In all the houses in which he lived the stairs attracted him first of all. And when I was observing him I had no doubts that the impressions of stairs would remain in him all his life and would be connected with all emotions of a strange, attractive and dangerous character.

Returning to the methods of my observations, I must note a curious fact demonstrating that dreams change by the fact of their being observed, namely that several times I *dreamed* that I was observing my dreams. My original aim was to create consciousness in dreams, i.e. to attain the capacity of realising in sleep that I was sleeping. In " half-dream states " this was there from the very beginning. As I have already said, I both slept and did not sleep at the same time. But soon there began to appear " false observations," i.e. merely new dreams. I remember once seeing myself in a large empty room without windows. Besides myself there was in the room only a small black kitten. " I am dreaming ", I say to myself. " How can I know whether I am really asleep or not ? Suppose I try this way. Let this black kitten be transformed into a large white dog. In a waking state it is impossible and if it comes off it will mean that I am asleep." I say this to myself and immediately the black kitten becomes transformed into a large white dog. At the same time the opposite wall disappears, disclosing a mountain landscape with a river like a ribbon receding into the distance.

" This is curious ", I say to myself ; " I did not order this landscape. Where did it come from ? " Some faint recollection begins to stir in me, a recollection of having seen this landscape somewhere and of its being somehow connected with the white dog. But I feel that if I let myself go into it I shall forget the most important thing that I have to remember, namely, *that I am asleep and am conscious of myself*, i.e. that I am in the state for which I have long wished and which I have been trying to attain. I make an effort not to think about the landscape, but at that moment some power seems to drag me *backwards*. I fly swiftly through the back wall of the room and go on flying in a straight line, all the time backwards and with a terrible noise in my ears, suddenly come to a stop and awake.

The description of this *backward* flying and the accompanying noise can be found in occult literature, where some special meaning is ascribed to them. But in reality there is no meaning in them except probably that of an inconvenient position of the head or slightly deranged circulation of the blood.

It was in this way, *flying backwards*, that people used to return from the witches' Sabbath.

And speaking generally, false observations, i.e. dreams within dreams, must have played a great part in the history of " magic ", miraculous transformations, etc.

False observations like the one described occurred several times, remained in my memory very vividly and helped me very much in elucidating the general mechanism of sleep and dreams.

I wish now to say a few words on this general mechanism of sleep.

First, it is necessary to understand clearly that sleep may be of different degrees, of different depths. We can be more asleep or less asleep, nearer to the possibility of awaking or further from the possibility of awaking. We usually remember only those dreams which we have when near to the possibility of awakening. Dreams which we have in deep sleep, i.e. far from the possibility of awaking, we do not remember at all. People who say they do not remember dreams sleep very soundly. People who remember all their dreams or at any rate many of them are really only half asleep. The whole time they are near the possibility of awaking. And, as a certain part of the inner instinctive work of our organism is best performed in deep sleep and cannot be well carried out when a man is only half asleep, it is obvious that the absence of deep sleep weakens the organism, prevents it renewing its spent forces and eliminating the used-up substances, and so on. The organism does not rest sufficiently. As a result it cannot produce sufficiently good work, is sooner worn out, more easily falls ill. In a word, deep sleep, that is, sleep without dreams, is in all respects more useful than sleep with dreams. And the experimenters who encourage people to remember their dreams render them a truly bad service. The less a man remembers his dreams the more soundly he sleeps and the better it is for him.

Further, it is necessary to note that we make a very great mistake when we speak about the creation of mental pictures in sleep.

Thus, we speak only of the head, brain thinking, and we ascribe to it the chief part of the work of creating dreams as well as all our thinking. This is utterly wrong. Our legs also think, think quite independently of and quite differently from the head. Arms also think : they have their own memory, their own mental images, their own associations. The back thinks, the stomach thinks, each part of the body thinks independently. Not one of these thinking processes reaches our consciousness in a waking state, when the head-thinking, operating chiefly by words and visual images, dominates everything else. But when the head-consciousness calms down and becomes clouded in the state of sleep, especially in the deeper forms of sleep, immediately other consciousnesses begin to speak, namely those of feet, hands, fingers, stomach, those of other organs, of various groups of muscles. These separate consciousnesses in us possess their own conceptions of many things and phenomena, for which we sometimes have also head-conceptions and sometimes have not. This is precisely what most prevents us understanding our dreams. In sleep the mental images which belong to the legs, arms, nose, tips of the fingers, to the various groups of motor muscles, become mixed with our ordinary verbal-visual images. We have no words and no forms for the expression of conceptions of one kind in conceptions of another kind. The visual-verbal part of our psychic apparatus cannot remember all these utterly incomprehensible and foreign images. In our dreams, however, these images play the same rôle as the visual-verbal images, if not a greater one.

The following two reservations that I make here should be remembered in every attempt at the description and classification of dreams. The first is that there are different states of sleep. We can only catch the dreams which pass near the surface ; as soon as they go deeper, we lose them. And the second is that no matter how we try to remember and exactly describe our dreams, we remember and describe only *head-dreams*, i.e. dreams consisting of visual-verbal images ; all the rest, i.e. the enormous majority of dreams, will escape us.

To this must be added another circumstance of very great importance. In sleep the head-consciousness itself changes. This means that man cannot in sleep think about himself *unless the thought is itself a dream*. A man can never pronounce his own name in sleep.

If I pronounced my name in sleep, I immediately woke up. And I understood that we do not realise that the knowledge of one's name for oneself is already a different degree of consciousness as compared with sleep. In sleep we are not aware of our own existence, we do not separate ourselves from the general picture which moves around us, but we, so to speak, move with it. Our "I" feeling is much more obscured in sleep than in a waking state. This is really the chief psychological feature which determines the state of sleep and expresses the whole difference between sleep and the waking state.

As I pointed out above, observation of dreams very soon brought me to the necessity for classification. I became convinced that our dreams differ very greatly in their nature. The general name of "dreams" confuses us. In reality dreams differ from one another as much as things and events which we see in a waking state. It would be quite insufficient to speak simply about "things", including in this planets, children's toys, prime ministers and paintings of the palæolithic period. This is exactly what we do in relation to "dreams". This certainly makes the understanding of dreams practically impossible and creates many false theories, because it is equally impossible to explain different categories of dreams on the basis of one common principle, as it would be prime ministers and paleolithic paintings.

Most of our dreams are entirely accidental, entirely chaotic, un-connected with anything and *meaningless*. These dreams depend on accidental associations. There is no consecutiveness in them, no direction, no idea.

I will describe one such dream, which was noted in *a half-dream state*.

I am falling asleep. Golden dots, sparks and tiny stars appear and disappear before my eyes. These sparks and stars gradually merge into a golden net with diagonal meshes which moves slowly and regularly in rhythm with the beating of my heart, which I feel quite distinctly. The next moment the golden net is transformed into rows of brass helmets belonging to Roman soldiers marching along the street below. I hear their measured tread and watch them from the window of a high house in Galata, in Constantinople, in a narrow lane, one end of which leads to the old wharf and the Golden Horn with its ships and steamers and the minarets of Stamboul behind them. The Roman soldiers march on and on in close ranks along the lane. I hear their heavy measured tread, and see the sun shining on their helmets. Then suddenly I detach myself from the window-sill on which I am lying, and in the same reclining position fly slowly over the lane, over the houses, and then over the Golden Horn in the direction of Stamboul. I smell the sea, feel the wind, the warm sun. This flying gives me a wonderfully pleasant sensation, and I cannot help opening my eyes.

This is a typical dream of the first category, i.e. of dreams which depend on accidental associations. Looking for a meaning in these dreams is exactly the same as telling fortunes by coffee grounds.

The whole of this dream passed before me when in a "half-dream state". From the first moment to the last I observed how pictures appeared and how they were transformed into one another. The golden sparks and dots were transformed into a net with regular meshes. Then the golden net was transformed into the helmets of the Roman soldiers. The pulsation which I heard was transformed into the measured tread of the marching detachment. The sensation of this pulsation means the relaxation of many small muscles, which in its turn produces a sensation of slight giddiness. This sensation of slight giddiness was immediately manifested in my seeing the soldiers, while lying on the window-sill of a *high* house and looking down; and when this giddiness increased a little, I rose from the window and flew over the gulf. This at once brought with it by association the sensation of the sea, the wind and the sun, and if I had not awakened, probably at the next moment of the dream I should have seen myself in the open sea, on a ship, and so on.

These dreams are sometimes remarkable for a particular absurdity, for quite impossible combinations and associations.

I remember one dream, in which for some reason a very great part was played by a large number of geese. Then somebody asks : " Would you like to see a *gosling*? you have certainly never seen a gosling." And at this moment I agree that I have never seen goslings. Next moment they bring me on an orange silk cushion a very strange-looking sleeping grey kitten, twice as long and thin as an ordinary kitten. And with great interest I examine the *gosling* and say that I never thought they were so strange.

If we place those dreams of which I have now spoken, that is, chaotic or incoherent, in the first category, we must place in the second category dramatic or invented dreams. Usually these two categories are intermixed, that is, an element of invention and fantasy enters into chaotic dreams, while invented dreams contain many accidental associations, images and scenes, which very often completely change their original direction. Dreams of the second category are the easiest to remember, for they are most like ordinary day-dreaming.

In these dreams a man sees himself in all kinds of dramatic situations. He travels in various distant lands, fights in wars, saves himself from some danger, chases somebody, sees himself surrounded by a crowd of people, meets all his friends and acquaintances dead and alive, sees himself at different periods of his life ; though grown-up he sees himself at school, and so on.

It happens that some dreams of this kind are very interesting in their technique. They contain a quantity of such subtle material of observation, memory and imagination as man does not possess while awake. This is the first thing that struck me in dreams of this kind when I began to understand something about them.

If I saw in my dream one of my friends whom perhaps I had not seen for several years, he spoke to me in his own language, in his own voice, with his own intonations and inflections, with his own characteristic gestures ; and he said precisely what only he could say.

Every man has his own manner of expressing himself, his own manner of thinking, his own manner of reacting to outward phenomena. No man can speak or act for another. And what first attracted my attention in these dreams was their wonderful artistic exactitude. The style of each man was kept throughout to the smallest detail. It happened that certain features were exaggerated or expressed symbolically. But there was never anything incorrect, anything inconsistent with the type.

In dreams of such a kind it happened that I saw more than once ten or twenty people simultaneously whom I had known at different periods of my life, and in not one of them was there ever the slightest mistake or the slightest inexactitude.

This was something more than memory ; it was artistic creation, because it was quite clear to me that many details which had obviously gone from my memory were reconstructed, so to speak, on the spot, and they corresponded completely to what ought actually to have been there.

Other dreams of this kind surprised me by their thoroughly thought out and elaborated plan. They had a clear and well-conceived plot which was unknown to me beforehand. All the dramatis personæ appeared at the right moment and said and did everything they had to do and say in conformity with the plot. The action could take place and develop in the most varied conditions, could be transferred from the town to the country, to lands unknown to me, to the sea; the strangest types could enter into these dramas. I remember, for instance, one dream, full of movement, dramatic situations and the most varied emotions. If I am not mistaken it was during the Japanese war. In the dream it was a war in Russia itself. A part of Russia was occupied by the armies of some strange people, called by a strange name, which I have forgotten. I had to pass at all costs through the enemy lines on some extremely important personal affairs. In connection with this a whole series of tragic, amusing, melodramatic incidents occurred. All this would have made a complete scenario for cinema production; and everything was in its right place, nothing was out of tune with the general course of the play. There were many interesting types and scenes. The monk with whom I spoke in a monastery still lives in my memory ; he was entirely outside life and outside all that took place around him, and at the same time he was full of little cares and little anxieties connected at that moment with me. The strange colonel of the enemy army with a pointed grey beard and incessantly blinking eyes was fully a living man and at the same time a very clear and definite type of man-machine, whose life is divided into several compartments with impenetrable partitions. Even the type of his imaginary nationality, the sound of the language he spoke with other officers, all this was in perfect keeping. The dream was full of small realistic details. I galloped through the enemy lines on a big white horse, and during one of the halts I brushed some white hairs off my coat with my sleeve.

I remember that this dream interested me very much because it showed me quite clearly that there was in me an artist, sometimes very naïve, sometimes very subtle, who worked at these dreams and created them out of the material which I possessed but could never use in full measure while awake. And I saw that this artist was extraordinarily versatile in his knowledge, capacities and talents. He was a playwright, a producer, a scene-painter, and a remarkable *actor-impersonator*. This last capacity in him was possibly the most astonishing of all. It especially struck me because I had very little of this capacity when awake. I never could imitate people, never could reproduce their voices, intonations, gestures, movements ; never could repeat the most characteristic words or phrases even of the people most familiar to me ; in the same way I never could reconstruct accents and peculiarities of speech. But I could do all this in dreams. The striking capacity for impersonation which manifested itself in dreams would undoubtedly have been a great talent had I been able to make use of it when awake. And I understood that this was not peculiar to me alone. This capacity for impersonation, for dramatisation, for arranging the picture, for stylisation, for symbolisation, lies within every man and is manifested in his dreams.

Dreams in which people see their dead friends or relations strike their imagination so strongly because of this remarkable capacity for impersonation inherent in themselves. This capacity can sometimes function in a waking state when man is absorbed in himself or separates himself from the immediate influences of life, and from usual associations.

After my observations of impersonation in dreams I entirely ceased to be surprised at tales of spiritualistic phenomena, of voices of people long dead, of " communications " and advice coming from them, etc. It can even be admitted that by following this advice people have found lost things, bundles of letters, old wills, family jewels or buried treasures. Certainly the majority of such tales are pure invention, but sometimes, although possibly very seldom, such things happen, and in that case they are undoubtedly based on impersonation. Impersonation is an art, although unconscious, and art always contains a strong " magic " element; and the magic element implies new discoveries, new revelations. A true and exact impersonation of a man long since dead can be magic like this. The impersonated image not only can say in this case what the man who reproduces it knows consciously or subconsciously, that is, without accounting for it to himself, but it can say definitely even things such as the man does not know, things which follow from the very nature of its being, from the nature of its life, that is, something that actually happened and that only it could know.

My own observation of impersonation did not go beyond observing the reproduction of what I had once known, heard or seen, with very small additions.

I remember two cases which explained to me a great deal in relation both to the origin of dreams and to " spiritualistic communications " from the world beyond. It happened after the time when I was occupied with the problem of dreams, on the way to India. I was alone. My friend S., with whom I had travelled in the East previously and with whom I had planned to go to India, had died a year before, and involuntarily, especially at the beginning of the journey, I thought about him and felt his absence.

And it happened twice—once on a boat in the North Sea and a second time in India,—that I distinctly heard his voice, as though he was entering my mental conversation with myself. On both occasions he spoke in the manner in which he alone could speak and said what he alone could say. Everything, his style, his intonation, his manner of speech, his way with me, all was in these few sentences.

Both times it happened on quite unimportant occasions, both times he joked with me in his usual manner. Of course I never thought for a moment that there could be anything " spiritualistic " in it; obviously he was in me, in my memory of him, and something within me reproduced him, " impersonated " him in these moments.

This kind of impersonation sometimes occurs in mental conversations with absent friends. And in these mental conversations, exactly as people who are dead can do, they can tell us things which we do not know.

In the case of people who are alive such incidents are explained by telepathy; in the case of the dead, by their existence after death and the possibility of their entering into telepathic communications with those alive.

This is the way things are usually explained in spiritualistic works. It is very interesting to read these spiritualistic books from the point of view of the study of dreams. I could distinguish different categories of dreams in the spiritualistic phenomena described : unconscious and chaotic dreams, invented dreams, dramatic dreams and one more, a very important category, which I would call imitative. This imitative category is curious in many respects, because although in many cases the material of these dreams is quite clear in our waking state, we should not be able to use it so skilfully as we do when asleep. Here again " the artist " is at work. Sometimes he is a producer, sometimes a translator, sometimes *an obvious plagiarist* changing in his own way and ascribing to himself what he has read or heard.

The phenomena of *impersonation* have also been described in scientific literature on the study of spiritualism. F. Podmore in his book *Modern Spiritualism* (London, 1902, Vol. II, pp. 302–303), cites an interesting case from *The Proceedings of the Society for Psychical Research* (Vol. XI, pp. 309–316).

Mr. C. H. Tout, principal of Buckland College, Vancouver, describes his experiences at spiritualistic séances. During these séances some persons were afflicted with spasmodic twitchings in their hands and arms and with other involuntary movements. Tout himself in these cases felt a strong impulse to imitate these movements.

At later séances he on several occasions yielded to similar impulses to assume a foreign personality. In this way he acted the part of a deceased woman, the mother of a friend then present. He put his arm round his friend and caressed him, as his mother might have done, and the personation was recognised by the spectators as a genuine case of " spirit control ".

On another occasion Mr. Tout, having under the influence of music given various impersonations, was finally oppressed by a feeling of coldness and loneliness, as of a recently disembodied spirit. His wretchedness and misery were terrible, and he was only kept from falling to the floor by some of the other sitters. At this point one of the sitters made the remark, which I remember to have overheard, " It is father controlling him ", and I then seemed to realise who I was and whom I was seeking. I began to be distressed in my lungs, and should have fallen if they had not held me by the hands and let me back gently upon the floor. As my head sank back on the carpet I experienced dreadful distress in my lungs and could not breathe. I made signs to them to put something under my head. They immediately put the sofa cushions under me, but this was not sufficient —I was not raised high enough yet to breathe easily—and they then added a pillow. I have the most distinct recollection of a sigh of relief I now gave as I sank back like a sick, weak person upon the cool pillow. I was in a measure still conscious of my actions, though not of my surroundings, and I have a clear memory of seeing myself in the character of my dying father lying in the bed and in the room

in which he died. It was a most curious sensation. I saw his shrunken
hands and face, and lived again through his dying moments; only
now I was both myself—in some indistinct sort of way—and my father,
with his feelings and appearance.

I remember a curious case of this category of pseudo-authorship.
It must have been about thirty years ago.

I awoke with a clear memory of a long and, as it seemed to me,
very interesting story, which I thought I had written in my dreams.
I remembered it in every detail and decided to write it down at the
first free moment, first as a specimen of " creative " dreams, second,
thinking that I might use the theme some day, although the story
had nothing in common with my usual writings and entirely differed
from them in type and character. But about two hours later, when
I began to write down the story, I noticed in it something very familiar
and suddenly, to my great amazement, I saw that it was a story by
Paul Bourget, which I had read not long before. The story was
altered in a curious way. The action which in Bourget's book
unfolded from one end, started in my dream from the other end.
The action took place in Russia, all the characters had Russian names,
and a new person was added introducing a definitely Russian atmosphere. I rather regret now that I did not write down the story at
the time as I constructed it in my dream. It undoubtedly contained
much of interest. First of all there was the extraordinary quickness
of the work. In normal conditions, when awake, such a turning
inside out of somebody else's story of similar length, transplanting
the action into another country and adding a new person who appears
in almost every scene, would require, according to my estimate, at
least a week's work. In sleep, however, it was done without any
expenditure of time, simply in the course of the progress of the action.

This extraordinary speed of mental work in sleep has many times
attracted the attention of investigators, and their observations have
given rise to many wrong deductions.

There is a well-known dream, much quoted but never fully understood, which is described by Maury in his book *Sleep and Dreams*,
which in his opinion establishes that one moment is sufficient for a
very long dream.

I was slightly indisposed and was lying in my room; my mother
was near my bed. I am dreaming of the Terror. I am present at
scenes of massacre; I appear before the Revolutionary Tribunal; I
see Robespierre, Marat, Fouquier-Tinville, all the most villainous
figures of this terrible epoch; I argue with them; at last, after many
events which I remember only vaguely, I am judged, condemned to
death, taken in a cart, amidst an enormous crowd, to the square of the
Revolution; I ascend the scaffold; the executioner binds me to the
fatal board, he pushes it, the knife falls; I feel my head being severed

from the body; I awake seized by the most violent terror, and I feel on my neck the rod of my bed which had become suddenly detached and had fallen on my neck as would the knife of the guillotine. This happened in one instant, as my mother confirmed to me, and yet it was this external sensation that was taken by me for the starting point of the dream with a whole series of successive incidents. At the moment that I was struck the memory of the terrible machine, the effect of which was so well reproduced by the rod of the bed's canopy, had awakened in me all the images of that epoch of which the guillotine was the symbol.[1]

Maury explained his dream by the extraordinary speed of the work of imagination in sleep, and it followed from his explanations that in some tenth or hundredth parts of a second, which passed between the moment when the bar struck his neck and his awakening, he constructed the whole dream, which was full of movement and dramatic effect, and seemed to last a long time.

But Maury's explanation is not sufficient and is wrong in its essence. It overlooks one most important circumstance. In reality the dream took a little longer than Maury thought, possibly several seconds, a fairly long period of time for a mental process; whereas for his mother his awakening might have appeared instantaneous or *very quick*.

What happened in reality was as follows. The fall of the rod brought Maury into a " half-dream state ". In this " half-dream state " the chief feeling was fear. He was afraid to wake up, afraid to explain to himself what had happened to him. The whole of his dream is created by this question : what has happened to me? This suspense, the uncertainty, the gradual disappearance of hope, are very well rendered in his dream as he tells it.

But there is one more very characteristic feature in Maury's dream which he did not notice. This is that events in his dream followed not in the order which he describes, but *from the end towards the beginning*.

This often happens in invented dreams, and it is one of the curious qualities of dreams, which may even have been noted somewhere in special literature on the subject. Unfortunately the importance and meaning of this quality have not been pointed out and the idea has not entered the usage of ordinary thought, though this capacity of dreams to develop backwards explains a great deal.

The backward development of dreams means that when we awake, we awake at the moment of the *beginning* of the dream and remember it as starting from this moment, that is, in the normal succession of events. Maury's first impression was : Oh God, what has happened

[1] *Le sommeil et les rêves, études psychologiques sur ces phénomènes*, by L. F. Alfred Maury, Paris, Didier et Cie, éditeurs, 1861, pp. 133–134.

to me ? Answer: I am guillotined. Imagination at once draws the picture of the execution, the scaffold, the guillotine, the executioner. At the same time the question arises : how can it all have happened ? How can I have got on to the scaffold ? In answer there again come pictures of the Paris streets, of the crowds of the time of the Revolution, of the tumbril in which the condemned were driven to the scaffold. Then again a question, with the same anguish wringing the heart and with the same feeling that something terrible and irreparable has happened. And in answer to these questions there appear pictures of the Tribunal, the figures of Robespierre, Marat, scenes of massacre, general pictures of the Terror, explaining all that happened. At this moment Maury awoke, that means, he opened his eyes. In reality he awoke long ago, possibly several seconds before. But having opened his eyes and remembering the last moment of the dream, the scenes of the Terror and massacre, he began at once to reconstruct the dream in his mind, starting from that moment. The dream began to unfold before him in the normal order, from the beginning of events to the end, from the scene at the tribunal to the fall of the knife of the guillotine, or, in reality, to the fall of the rod.

Later when writing down or telling his dream he never doubted for a second that he actually had the dream in this order, that is to say, he never imagined the possibility of dreaming a dream in one order of events and remembering it in another. Another problem arose therefore before him : how such a long and complex dream could flash past in one moment, for he was certain that he awoke at once (he did not remember the " half-dream state "). This he explained by the extraordinary swiftness of the development of dreams, whereas in reality the explanation requires the understanding first of " half-dream states " and second of the fact that dreams can develop in *reverse order*, from end to beginning, and be remembered in the *right order*, from beginning to end.

The development of dreams from end to beginning happens fairly often, but of course we always remember these dreams in the normal order because they end with the moment from which they would begin in the normal development of events, but are remembered or imagined from this moment.

The emotional states in which we may be during sleep often produce very curious dreams. They colour with one shade or another the usual half-chaotic, half-invented dreams, make them wonderfully alive and real, and cause us to seek in them a deep meaning and significance.

I will cite here one dream which undoubtedly could be interpreted spiritualistically, though of course there is no spiritualism in it (I had this dream when I was seventeen or eighteen). I dreamed of Lermontoff. I do not remember the visual image, but he told me in a strange hollow and strangled voice that he did not die when he was thought to have been killed. " I was saved," he said, slowly and in a low voice. " My friends arranged it. The Circassian who jumped into the grave and knocked off the earth with his dagger, pretending that it was necessary to help the coffin to pass. . . . It was connected with that. At night they dug me out. I went abroad and lived there for a long time, only I did not write anything more. No one knew about it except my sisters. Later I really died."

I awoke from this dream in an unusually depressed state. I was lying on my left side, my heart was beating fast, and I was feeling inexpressible anguish. This anguish was really the chief motive which, in connection with accidental images and associations, created the whole dream. So far as I can remember, my first impression of " Lermontoff " was the hollow strangled voice, full of some peculiar sadness. Why I replied to myself that it was Lermontoff it is difficult to say. It is possible that there was in this an emotional association. Very likely the description of the death and burial of Lermontoff might have produced a similar impression on me at one time. Lermontoff's saying that he did not die, that he was buried alive, accentuated this emotional tone still more. A curious feature of this dream was the attempt to connect the dream with facts. In the description of Lermontoff's burial in some biographies, it is stated, on the strength of the accounts of eye-witnesses, that the coffin could not pass into the recess at the side of the grave and that a mountaineer jumped down and knocked off the earth with his dagger. In my dream something was connected with this incident. Then " Lermontoff's sisters ", who alone knew that he was alive. I thought even in my dream that he said " sisters " meaning " cousins ", as though for some reason or other he did not wish to speak clearly. All this followed from the chief motive of the dream, a feeling of depression and mystery.

There is no doubt that this dream would have been interpreted by spiritualists in a spiritualistic sense. Speaking generally, the study of dreams is the study of " spiritualism ", because " spiritualism " draws all its contents from dreams. And as I pointed out before, spiritualistic literature gave me very interesting material for the explanation of dreams.

But apart from this, spiritualistic literature undoubtedly creates

whole series of " spiritualistic " dreams, just as the cinematograph
or detective novels undoubtedly play a very important part in the
creation of dreams.

Modern attempts at the investigation of dreams as a rule hardly
take into consideration the character of a man's reading and still less
his favourite amusements like theatres, cinemas, races, etc., whereas
it is precisely from these that the chief material of dreams comes,
especially in the case of people whose everyday life contains but
few impressions.   It is reading and spectacular sights that create allegorical, symbolical and similar dreams.   The rôle played by advertise-
ments and posters in creating dreams is also quite disregarded.

The building up of visual images is sometimes very strange in
dreams.   I have already mentioned the fact that dreams are principally
built according to associations of impressions and not according to
associations of facts.   And, for instance, in visual images entirely
different people, with whom we come into contact at entirely different
periods of our lives, very often become merged and united into one
person.

A young girl, a political prisoner who spent a long time in the
Boutirsky prison in Moscow (in 1906–1908), told me during my
visits, from behind two rows of bars, that in her dreams the impressions of the prison were completely mixed up with the impressions
of the " Institute " [1] which she had left only six years before.   In
her dreams the prison warders became confused with former " class-
ladies " and " inspectresses " (house-mistresses).   Summonses before
the prosecutor and cross-examination were lessons, the coming trial
was the final examination, and everything was similarly confused.

In this case the connecting link was undoubtedly the similarity
of emotional experiences, the boredom, the continual constraint and
the general absurdity of all the surroundings.

Another dream has remained in my memory, this time merely
an amusing one, in which was manifested the principle of the personi-
fication of ideas opposite to the one described.

Long ago when I was quite young I had a friend in Moscow who
accepted a situation in the south of Russia and went there.   I remember
seeing him off at the Kursk railway station.

About ten years later I saw him in my dream.   We were sitting
at a table in the station restaurant drinking beer exactly as we had
done when I saw him off.   But *we were three* : I, my friend as I
remembered him, and my friend as he probably must have become in
some part of my mental picture of him, a stout middle-aged man much

---

[1] A privileged government school for girls, of the type established in Russia in the
18th century and having the character of French convents.

older than he could have been in reality, dressed in an overcoat with a fur collar and having slow and assured movements. As usually happens in dreams this combination did not surprise me in the least, and I took it as though it was the most ordinary thing in the world.

I have now mentioned several categories of dreams, but these by no means cover all possible and existing categories. One of the reasons for the wrong interpretation of dreams is the inadequate understanding of the categories and a wrong division of dreams.

I have already pointed out that dreams differ among themselves not less than phenomena of the real world. All the examples given up to now relate to " simple " dreams, that is, to dreams which take place on the same level as our ordinary life, as our thinking and feeling in a waking state. But there are other categories of dreams. These dreams have their origin in the innermost recesses of life and rise high above the common level of our understanding and perception of things. These dreams can disclose a great deal that is unknown to us on the ordinary level of life, for instance, in showing us the future or the thoughts and feelings of other people or events unknown to us or remote from us. And they can also disclose to us the mysteries of being, show the laws governing life, bring us into contact with higher forces. These are very rare dreams, and one of the errors of the usual treatment of dreams is that these dreams are regarded as much more frequent than they are in actual fact. Their principles and ideas became to a certain extent comprehensible to me only after the experiments which I describe in the next chapter.

It must be understood that all that can be found about dreams in psychological literature refers to " simple " dreams. The confusion of ideas about these dreams depends, apart from wrong classification of the dreams themselves, to a considerable degree on wrong definition of the material of which dreams are made. Dreams are regarded as being created from *fresh* material, from the same material as that which goes to create the thoughts, feelings and emotions of our waking life. This is the reason why dreams in which a man performs actions or experiences emotions, which he could not have performed or experienced when awake, give rise to such multitudes of questions. The interpreters of dreams take it all quite seriously and create their own picture of a man's soul on the basis of these features. All this is of course quite wrong.

With the exception of dreams like those described in the beginning, such as the dream of the " quagmire " or " blindness ", which are created by sensations received during sleep, the chief material which goes to make up dreams is the refuse or used-up material of our psychic life.

It is the gravest mistake to think that ordinary dreams reveal us as we are somewhere in the unknown depths of our nature. To ourselves dreams cannot do this ; they picture either what has been and has gone by, or, still more often, what has not been and could not have been. Dreams are always a caricature, always a comic exaggeration, but an exaggeration which in most cases relates to some non-existent moment in the past or non-existent situation in the present.

The question is, what are the principles which create this caricature ? Why do dreams so contradict reality ? And here we meet with a principle which though not fully understood has nevertheless been noted in " psychoanalytical " literature. This is the principle of " compensation ". But the word itself is unsuccessful, and probably this unsuccessful word creates its own unsuccessful associations, which is the reason why the principle has never been wholly understood, but has on the contrary given rise to utterly wrong theories.

This idea of " compensation " has been connected with the idea of dissatisfaction. The action of the principle is understood in the sense that a man who is dissatisfied with something in life in regard either to himself or to others, compensates himself in dreams. A weak, unhappy, cowardly man sees himself brave, strong, attaining everything he desires. Some friend suffering from an incurable disease is seen by us in dreams as cured, full of strength and hope. Similarly, people who have had a long illness or have died in painful conditions appear to us in dreams, cured, content and happy. In this instance the interpretation is very near the truth, but nevertheless it is only half the truth.

In reality the principle is much wider, and the material of dreams is created not on the principle of compensation taken in a simple, psychological or life sense, but on the basis of what I would call the principle of *complementary tones* entirely without relation to our emotional feeling of those tones. This principle is very simple. If you look for some time at a red spot and then turn your eyes to a white wall, you will see a green spot. If you look for some time at a green spot and then take your eyes off you will see a red spot. Exactly the same thing happens in dreams. There exist for us no morals in dreams, *because* for good or bad our life is controlled by different moral rules. Every moment of our life is surrounded by different kinds of " thou shalt not ", and therefore " thou shalt not " does not exist in dreams. There exists for us nothing extraordinary in dreams, because in life we are astonished at every new or unusual combination of circumstances. There exists for us no law of the consecutiveness of phenomena in dreams, because this law governs everything in life, and so on.

The principle of complementary tones plays the chief rôle in our dreams, as much in those we remember as in those we do not remember; and without keeping this principle in view it is impossible to explain a whole series of dreams in which we do and apparently feel what we never do and never feel in life.

Very many things happen in dreams only because they never happen and never can happen in life. Dreams are very often the *negative* in relation to the *positive* of life. But again it should be remembered that this refers only to details. The composition of dreams is not the simple opposite of life, but an " opposite " turned inside out several times and in several senses. Therefore attempts to reconstruct from dreams the hidden causes of dreams are quite useless, and it is merely senseless to suppose that the hidden causes of dreams are the hidden motives of life in a waking state.

It remains for me to make a few remarks about the conclusions which resulted from my attempts to study dreams.

The more I observed dreams the wider became the field of my observations. At first I thought that we have dreams only in a definite state of sleep, near awakening. Later I became convinced that we have dreams all the time, from the moment we fall asleep to the moment we awake, but *remember* only the dreams near awakening. And still later I realised that we have dreams continuously, *both in sleep and in a waking state*. We never cease to have dreams, though we are not aware of this.

As the result of the above I came to the conclusion that dreams can be observed while awake. It is not at all necessary to be asleep in order to observe dreams. Dreams never stop. We do not notice them in a waking state, amidst the continuous flow of visual, auditory and other sensations, for the same reason for which we do not see stars in the light of the sun. But just as we can see the stars from the bottom of a deep well, so we can see the dreams which go on in us if, even for a short time, we isolate ourselves whether accidentally or intentionally, from the inflow of external impressions. It is not easy to explain how this is to be done. Concentration upon one idea cannot produce this isolation. An arrest of the current of usual thoughts and mental images is necessary. It is necessary to achieve for a short period " consciousness without thought ". When this consciousness comes dream images begin slowly to emerge through the usual sensations, and with astonishment you suddenly see yourself surrounded by a strange world of shadows, moods, conversations, sounds, pictures. And you understand then that this world is always in you, that it never disappears.

You come to a very clear although somewhat unexpected con- clusion : sleep and the waking state are not two states that *succeed one another*, or follow one upon another. The names themselves are incorrect. The two states are not *sleep* and *waking state*. They may be called *sleep* and *sleep plus waking state*. This means that when we awake sleep does not disappear, but to the state of sleep *there is added* the waking state, which muffles the voices of dreams and makes dream images invisible.

The observation of " dreams " in a waking state presents far fewer difficulties than observation in sleep and, moreover, observation in this case does not change their character, does not create new dreams.

After some experience, even the arresting of thoughts, the creation of consciousness without thought, becomes unnecessary. Dreams are always there. It is sufficient only to divide the attention, and you see how into the usual thoughts of the day, into the usual conver- sations, there enter thoughts, words, figures, faces, scenes, either from the past, from childhood, from school years, from travels, or from what has been read or heard at some time, or from that which has never happened but of which one was one day thinking or talking.

To the dreams observable only in a waking state belongs (in my case) the strange sensation which is known to many people and has many times been described, though it has never been fully explained— *the sensation that this has happened before.*

Suddenly in some *new* combination of circumstances, among new people, in a new place, a man stops and looks with astonishment about him—this has happened before ! But when ? He cannot say. Later he tells himself that it *could not be so*, he has never been here or in these surroundings, has never seen these people.

Sometimes it happens that these sensations are very persistent and long, sometimes very quick and elusive. The most interesting of them occur with children.

A distinct realisation that it has happened before is sometimes absent in these sensations. But it happens sometimes without any visible or explainable cause that some definite thing, a book, a toy, a dress, a certain face, a house, a landscape, a sound, a tune, a poem, a smell, strikes the imagination as something familiar, well known, touching upon the most hidden feelings, evokes whole series of vague and fleeting associations and remains in the memory for the whole lifetime.

With me these sensations (with a clear and distinct idea that this has happened before, that I have seen it before) began when I was about six years old. After eleven they became much rarer. One of them, extraordinary for its vividness and persistence, occurred when I was nineteen.

The same sensations, but without a clearly pronounced feeling of repetition, began still earlier, from very early childhood, and were particularly vivid during the years when the sensations of repetition appeared, that is, from six to eleven ; and they also came later from time to time in various conditions.

Usually when these sensations are treated of in psychological literature, only the first kind is meant, namely, the sensations with a clearly pronounced idea of repetition.

According to psychological theories, sensations of this kind are produced by two causes. Firstly, they depend on breaks in consciousness, when consciousness suddenly disappears for one quite imperceptible moment and then flashes out again. In this case the situation in which one finds oneself, that is, all that surrounds one, seems to one to have happened before, possibly long ago in the unknown past. The " breaks " themselves are explained by the possibility of the same psychic function being carried out by different parts of the thinking apparatus. As a result of this, one function having accidentally stopped in one part is immediately taken up and continued in another, producing the impression that the same situation has occurred some time previously. Secondly, the same sensation may be produced by an associative resemblance between totally different experiences, when a stone or a tree or any object may remind one of somebody one knew very well, or of some place, or of a certain incident in one's life. This happens when for instance one feature or line of a stone reminds you of some feature in a man or in another object ; this can also give the sensation that *this has happened before.*

Neither of these theories explains the reason why in most cases the sensation that *this has happened before* occurs chiefly in children and almost always disappears later. On the contrary, according to these theories, the sensations described should grow more frequent with age.

Both the above theories are deficient in that they do not explain *all* the existing facts of the sensation of repetition. Exact observations show *three categories* of such sensations. The first two categories are explained (although not fully) by the above psychological theories. The peculiarity of these two categories is that they usually occur in a partly clouded consciousness, almost in a half-dream state, although this may not be realised by the man himself.

The third category of sensations that this has happened before stands quite apart, and its peculiarity is that the *sensations of repetition* are connected in these cases with an especially clear waking state of consciousness and a heightened self-feeling.

I shall speak of these sensations and their meaning in another place.

In speaking of the study of dreams it is impossible to pass over another phenomenon, which is directly connected with it and which remains unexplained up to the present time, in spite of some possibility of experimenting with it.

I refer to *hypnotism.* The nature of hypnotism, i.e. its causes, and also the forces and laws that make it possible, remains unknown. All that can be done is to establish conditions in which phenomena of hypnotism may occur, and the possible limits, results and consequences of these phenomena.

In this connection it must be noted that the general reading public has attached to the word hypnotism such a number of wrong conceptions that before speaking of what is possible under the term hypnotism it must be made clear what is impossible.

Hypnotism in the popular and fantastic meaning of the word and hypnotism in the scientific or real meaning of the word are two entirely different ideas.

In the real meaning the content of all the facts united under the general name of hypnotism is very limited.

By being subjected to special kinds of treatment a man can be brought to a particular state, called the hypnotic state. Although there exists a school which asserts that any man can be hypnotised at any time, facts tell against this. In order to be hypnotised, to fall into a hypnotic state, a man must be perfectly passive, i.e. know that he is being hypnotised and not resist it. If he does not know, the ordinary course of thoughts and actions suffices to protect him from the possibility of hypnotic action. Children, drunken men, madmen, do not submit to hypnosis, or submit very badly.

There exist many forms and degrees of the hypnotic state. They can be created by various methods. Passes and strokings of a certain kind, which provoke relaxation of the muscles, a fixed gaze into the eyes, flashing mirrors, sudden impressions, a loud shout, monotonous music : all these are means of hypnotising. Besides this narcotics are used, although the use of narcotics in hypnosis has been very little studied, and description of their use is hard to find even in special literature on the subject. But narcotics are used far more often than is thought, and for two purposes : first for the weakening of the resistance to hypnotic action, and second for the strengthening of the capacity to hypnotise. There are narcotics which act differently on different people, and there are narcotics which have a more or less uniform action. Almost all professional hypnotists use morphia or cocaine in order to be able to hypnotise. Different narcotics are used also for the person hypnotised ; a weak dose of chloroform very much increases the capacity of a man to submit to hypnosis.

What actually occurs in a man when he is hypnotised and by what force another man hypnotises him, are questions which science cannot answer. All that we know up to now gives us the possibility of establishing only the external form of the hypnotic state and its results. The hypnotic state begins with simple weakening of the will. Control of ordinary consciousness and ordinary logic weakens. *But it never disappears altogether.* With skilful action, the hypnotic state is intensified. The man thus passes into a state of a particular kind ; the external side of this state is characterised by its resemblance to sleep (in deep states unconsciousness and even insensibility appear), and the internal side by an increase of suggestibility. The hypnotic state is therefore defined as the *state of maximum suggestibility.*

In itself hypnosis does not comprise any suggestion, and is possible without any suggestion, particularly if purely mechanical means are used, such as mirrors, etc. But suggestion may play a certain part in the creation of the hypnotic state, particularly in repeated hypnotising. This fact, and also in general the confusion of ideas as to the possible limits of hypnotic action, makes it very difficult for non-specialists (as well as for many specialists) to distinguish exactly between hypnosis and suggestion.

In actual fact they are two entirely different phenomena. Hypnosis is possible without suggestion, and suggestion is possible without hypnosis.

But if suggestion, whatever it be, takes place while the subject is in a hypnotic state, it will give notably greater results. There is no resistance, or almost none. A man can be made under hypnosis to do things which seem to him a complete absurdity, though only things which have no serious importance. It is equally possible to suggest to a man something for the future (post-hypnotic suggestion), i.e. it is possible to order some action, thought or feeling for a certain moment, on the following day or later. Then the man can be awakened. He will remember nothing. But at the appointed time, like a wound-up clockwork mechanism, he will do or at least will attempt to do what has been " suggested " to him. But again only up to a certain limit. It is impossible to make a man, when hypnotised or through post-hypnotic suggestion, do anything which would contradict his nature, tastes, habits, education, convictions or even merely his ordinary actions ; it is impossible to make him do anything which would provoke inner struggle in him. If such a struggle begins, the man *does not do* what has been suggested to him. The success of suggestion under hypnosis or of post-hypnotic suggestion consists precisely in suggesting to a man a series of *indifferent* actions which provoke in him no struggle. Suppositions that a man under hypnosis can be made to *know* something which he did not know in a normal state and which the hypnotiser does not know, or that a man under hypnosis can show a capacity for " clairvoyance ", that is, for knowing the future or seeing events occurring at a distance, are not confirmed by any facts. At the same time there are known many cases of unconscious suggestion on the part of the hypnotiser and a certain capacity for reading his thoughts on the part of the person hypnotised.

All that takes place in the mind of the hypnotiser, that is, the semi-conscious associations, imagination and anticipation of what according to him must happen, can be transferred to the person hypnotised by him. How the transference takes place it is impossible to establish, but the fact of this transference is very easy to prove if that which is known by the one is compared with that which is known by the other.

To this category are related phenomena of so-called " mediumism ".

There is a very curious book by a French author, de Rochas, who describes experiments with persons whom he hypnotised and made " remember " their previous " incarnations " on earth. In reading this book I was many times amazed that the author could avoid seeing that he himself was the creator of all these " incarnations ", *anticipating* what the hypnotised subject would say and *in this way suggesting to him what to say.*

This book gives very interesting material for the understanding of the process of the formation of dreams. It might have given even more important material for the study of the methods and forms of unconscious suggestion and unconscious thought-transference. But, unfortunately, the author, in his pursuit of fantastic " remembrances " of incarnations, did not see what was really valuable in his experiments and did not note many small details and particulars which would have given the possibility of reconstructing the processes of suggestion and transference of thoughts.

Hypnotism is applied in medicine as a means of action on the emotional nature of a man; for the struggle through suggestion with gloomy and depressed moods, with morbid fears and unhealthy tendencies and habits. And in those cases in which the pathological manifestations are not dependent on deep-seated physical causes, the use of hypnotism gives favourable results. However, with regard to these results, the opinions of specialists differ, and many assert that the use of hypnotism gives only short-lived useful results with a very strong reaction in the direction of the increase of undesirable tendencies, or, in the presence of seemingly favourable results, gives concomitant negative results, weakens the will and capacity of resistance to undesirable influences and makes a man even less stable than he was.

In general, hypnotism, in those cases in which the *psychical* nature of the patient is the object of action, stands on the level of a serious operation, and unfortunately is often applied without sufficient grounds and without sufficient understanding of the consequences of its use.

There exists another sphere in which hypnotism could be applied in medicine without any harm, namely the sphere of direct action (i.e. not through the mediation of the patient's psychical nature) on nerve centres, tissues, inner organs and inner processes. But unfortunately this sphere has been very little studied up to the present time.

Thus the limits of possible influence on a man with the aim of bringing him to a hypnotic state, as well as the limits of possible action on a man who is in a hypnotic state, are very well known and contain nothing enigmatic. The strengthening of the influence is possible only in the direction of strengthening the influence on the physical nature of man apart from his psychic apparatus. But it is precisely in this direction that attention has been turned least of all. On the contrary, current conceptions of hypnotism admit far greater possibilities of action on man's psychical nature than exist in actual fact.

There exist, for instance, very many stories about *mass hypnosis*, but all these stories, in spite of their wide circulation, are the purest invention, and most often are merely repetitions of similar stories which existed earlier.

In 1913 and 1914 I tried to find in India and Ceylon examples of mass hypnosis, with which, according to the descriptions of travellers, the performances of Indian jugglers or "fakirs" and some religious ceremonies are accompanied. But I did not succeed in seeing one single instance. Most of the performances, such, for example, as the raising of a plant from a seed ("mango trick") were mere tricks. And the often described "rope trick", in which a rope is thrown "up to the sky" and a boy climbs up it, etc., has obviously never existed, because not only did I not succeed in seeing it myself, but I never found a *single man* (European) who had seen it *himself*; they all knew of it only by what they had been told. A few educated Hindoos told me they had seen the " rope trick ", but I cannot accept their statements as credible because, besides a very fertile imagination, I noticed in them a strange reluctance to disappoint people who look in India for miracles.

I heard later that during the Prince of Wales' travels in India (in 1921 and 1922) the " rope trick " was sought for specially for him, but could not be found. In the same way India was searched for the " rope trick " for the Wembley exhibition of 1924, but it could not be found.

A man who knew India very well told me once that the only thing resembling the " rope trick " he had ever seen was some juggling by a Hindoo conjurer with a thin wooden hoop at the end of a long bamboo rod. The juggler made the hoop run up and down the rod. Possibly it is this that started the legend.

In the 2nd and 3rd issues of the *Revue Métapsychique* (Mars-Avril, Mai-Juin) for the year 1928 there is an article (by M. C. de Vesme) " La légende de l'hallucination collective à propos du tour de la corde pendue au ciel " The author gives a very interesting survey of the history of the " rope trick ", citing descriptions of the " rope trick " by eye-witnesses, stories told by people who had only heard about it, and the history of attempts to find and establish the real existence of this trick. Unfortunately, however, while denying the miraculous he himself makes several naïve assertions. For instance, he recognises the possibility of a " mechanical device concealed inside the rope ", which enables the rope to stand upright so that a boy can climb it. In another place he speaks of a photograph of the " rope trick ", in which one can distinguish a bamboo *inside the rope*.

Actually, if such a thing as a mechanical appliance inside the rope were possible it would be even more miraculous than the " rope trick " as it is usually described. I doubt whether even European technique could contrive such a device to be placed *inside* a thin and, presumably, fairly long rope, which would make the rope stand upright and allow a boy to climb it. But how a half-naked Hindoo juggler could have such a rope is totally incomprehensible. The " bamboo " inside the rope is still more interesting. The question arises here how the rope could be coiled if it had a bamboo inside it. Altogether the author of this very interesting survey of the study of Indian miracles has got, on this point, into a very strange position.

But stories of the miracles of fakirs make a necessary part of the descriptions of impressions of India and Ceylon. Not very long ago I happened to see a French book the author of which relates his adventures and experiences in Ceylon in recent years. To do him justice he caricatures everything he describes and makes no pretensions to seriousness. But he describes another " rope trick " in Kandy, this time with certain variations. Thus, the author, who was hidden on a verandah, was not hypnotised by the " fakir " and therefore did not see what his friends saw. Besides this, one of them photographed the whole of the performance with a cinematograph camera.

" *But when we developed the film* the same night,", writes the author, " there was nothing on it."

What is most amusing is that the author does not realise in what *the most miraculous* part of his last statement consists. But this persistence in the description of the " rope trick " and " mass hypnotism ", that is, precisely what does not exist, is very characteristic.

In speaking of hypnotism it is necessary to mention *self-hypnosis*.

The possibilities of self-hypnosis also are exaggerated. In reality self-hypnosis without the help of artificial means is possible only in a very feeble degree. By creating in himself a certain passive state a man can weaken the resistance which comes, for example, from logic or common sense, and surrender himself wholly to some desire. This is the possible form of self-hypnosis. But self-hypnosis never attains the forms of sleep or catalepsy. If a man seeks to overcome some great resistance in himself, he uses narcotics. Alcohol is one of the chief means of self-hypnosis. The rôle of alcohol, as a means of self-hypnosis, is still entirely unstudied.

Suggestion must be studied separately from hypnotism.

Hypnotism and suggestion are constantly confused ; the place therefore which they occupy in life is quite undetermined.

In reality, *suggestion* is the fundamental fact. Hypnotism might not exist in our life, nothing would be altered by this, but *suggestion* is one of the chief factors both in individual and in social life. If there were no suggestion, men's lives would have an entirely different form, thousands of the phenomena of the life surrounding us would be quite impossible.

Suggestion can be conscious and unconscious, intentional and unintentional. The sphere of conscious and intentional suggestion is extremely small in comparison with the sphere of unconscious and unintentional suggestion.

Man's *suggestibility*, i.e. his capacity to submit to surrounding suggestions, can be different. A man can be entirely dependent on suggestions, have nothing in himself but the results of suggestions and submit to all sufficiently strong suggestions, however contradictory they may be ; or he can show some resistance to suggestions, at least yield to suggestions only of certain definite kinds and repel others. But resistance to suggestion even of such a kind is a very rare phenomenon. Ordinarily a man is wholly dependent on suggestions ; and his whole inner make-up (and also his outer make-up) is entirely created and conditioned by prevailing suggestions.

From earliest childhood, from the moment of first conscious reception of external impressions, a man falls under the action of suggestions, intentional and unintentional. In this case certain feelings, rules, principles and habits are suggested to him intentionally ; and the ways of acting, thinking and feeling against these rules, principles and habits are suggested unintentionally.

This latter suggestion acts owing to the tendency to imitation which everyone possesses. People say one thing and do another. A child listens to one thing and imitates another.

The capacity for imitation in children and also in grown-up people greatly increases their suggestibility.

The dual character of suggestions gradually develops duality in man himself. From very early years he learns to remember that he must show the feelings and thoughts demanded of him at the given moment and never show what he really thinks and feels. This habit becomes his second nature. As time passes, he begins, also through imitation, to trust alike the two opposite sides in himself which have developed under the influence of opposite suggestions. But their contradictions do not trouble him, first because he can never see them together, and second because the capacity not to be troubled by these contradictions is *suggested* to him *because nobody ever is troubled.*

Home-education, the family, elder brothers and sisters, parents, relatives, servants, friends, school, games, reading, the theatre, newspapers, conversations, further education, work, women (or men), *fashion,* art, music, the cinema, sport, the jargon accepted in his circle, the accepted wit, obligatory amusements, obligatory tastes and obligatory taboos—all these and many other things are the source of new and ever new suggestions. All these suggestions are invariably dual, i.e. they create simultaneously what must be shown and what must be hidden.

It is impossible even to imagine a man free from suggestions, who really thinks, feels and acts as he himself can think, feel and act. In his beliefs, in his views, in his convictions, in his ideas, in his feelings, in his tastes, in what he likes, in what he dislikes, in every movement and in every thought, a man is bound by a thousand suggestions, to which he submits, even without noticing them, *suggesting to himself* that it is he himself who thinks in this way and feels in this way.

This submission to external influences so far permeates the whole life of a man, and his suggestibility is so great, that his ordinary, normal state can be called *semi-hypnotic*. And we know very well that at certain moments and in certain situations a man's suggestibility can increase still more and he can reach complete loss of any independent decision or choice whatever. This is particularly clearly seen in the psychology of a crowd, in mass movements of various kinds, in religious, revolutionary, patriotic or panic moods, when the seeming independence of the individual man completely disappears.

All this taken together constitutes one side of the " life of suggestion " in a man. The other side lies in himself and consists, first, in the submission of his so-called " conscious ", i.e. intellectual-emotional, functions to influences and suggestions coming from the so-called " unconscious " (i.e. unperceived by the mind) voices of the body, the countless obscure consciousnesses of the inner organs and inner lives ; and second, in the submission of all these inner lives to the completely unconscious and unintentional suggestions of the reason and the emotions.

The first, i.e. the submission of the intellectual-emotional functions to the instinctive, has been more elaborated in psychological literature ; though the greater part of what is written on these subjects must be taken very cautiously. The second, i.e. the submission of the inner functions to the *unconscious* influences of the nerve-brain apparatus, has been very little studied. Meanwhile, this last side offers enormous interest from the point of view of the understanding of suggestion and suggestibility in general.

A man consists of a countless number of lives. Each part of the body which has a definite function, each organ, each tissue, each cell, has its separate life and its own separate consciousness. These consciousnesses differ very greatly in their content and in their functions from the intellectual-emotional consciousness which is known to us and which belongs to the whole organism. But this last consciousness is by no means the only one. It is not even the strongest or the clearest. Solely by virtue of its position, so to say, on the border of the inner and outer worlds it receives predominant significance and the possibility of suggesting very many ideas to the obscure inner consciousnesses. The inner consciousnesses are constantly listening to the voice of reason and of the emotions. This voice attracts them, subjugates them to its power. Why ? It may seem strange, seeing that the inner consciousnesses are often more subtle and keen than the brain-consciousness. It is true that they are more subtle and keen, but they live in the dark, within the organism. The brain-consciousness appears to them as knowing more than they, as it is turned to the outer world. And the whole crowd of obscure inner consciousnesses incessantly follows the life of the outer consciousness and strives to imitate it. The head-consciousness is entirely ignorant of this and gives them thousands of different suggestions, which are very often contradictory, absurd and harmful to the organism.

The inner consciousnesses are a provincial crowd listening to the opinions of inhabitants of the capital, following their tastes, imitating their manners. What the " mind " and " feeling " say, what they do, what they wish, what they fear, becomes instantly known in the most distant, in the darkest, corners of the organism, and of course it is interpreted and understood in each of them in a different way. A perfectly casual, paradoxical idea of the brain-consciousness, which " comes into the head " casually and is forgotten casually, is taken as a revelation by some " connective tissue ", which of course remodels it in its own way and begins to " live " in conformity with this idea. The stomach can be entirely hypnotised by certain absurd tastes and aversions of a purely " æsthetic " character ; heart, liver, kidneys, nerves, muscles, may all in this or some other way submit to suggestions which are unconsciously given to them by thoughts and emotions. A considerable number of the phenomena of our inner life, particularly of undesirable phenomena, is in reality dependent on these suggestions. The existence and character of these obscure consciousnesses also explain a great deal in the world of dreams.

The mind and feeling forget or know nothing about this crowd which listens to their voices, and they often talk too loud when it would be better for them to be silent or not to express their opinions, since sometimes their opinions, unimportant and transient for themselves, may produce a very strong impression on the inner consciousnesses. If we do not wish to be in the power of unconscious self-suggestions, we must be careful of the words we use when we speak to ourselves and of the intonations with which we pronounce these words, although consciously we do not attach importance to these words and intonations. We must remember about all these obscure people, listening at the doors of our consciousness, drawing their own conclusions from what they hear, submitting with incredible ease to temptations and fears of every kind and starting to rush about in panic at some simple thought, that we may miss the train or lose a key. We must learn to consider the importance of these inner panics, or, for example, of the terrible depression that suddenly seizes us at the sight of a grey sky and rain beginning This means that the inner consciousnesses have caught a casual phrase : " What nasty weather ", which was said with great feeling, and they have understood it in their own way, that now the weather will always be nasty, that there is no way out and that it is not worth while living or working any longer.

But all this refers to unconscious self-suggestion. The limits of voluntary self-suggestion in our ordinary state are so insignificant that it is impossible to speak of any practical application of this force. Yet against all facts the idea of *self-suggestion* inspires confidence. And at the same time the study of involuntary suggestions and of involuntary suggestibility can never be popular because, more than anything else can do, it destroys millions of illusions and shows a man what he really is. And a man in no case wishes to know this, and he does not wish it because against it there acts the strongest suggestion existing in life, the suggestion which persuades a man to be and to appear other than he is.

1905–1929.

Globe Press Books

Globe Press Books publishes works of philosophical, psychological and spiritual importance. Some of these explain and expand the powerful psychological system originated by G.I. Gurdjieff. You can stay informed about our forthcoming publications by simply mailing us the Order Form on the last page of this book. Some of our other titles are detailed below. These books are available at fine bookstores everywhere, or they can be ordered directly from the publisher using the Order Form.

Body Types

by Joel Friedlander. *The Enneagram of Essence Types*. Learn how to recognize the physical and psychological tendencies of each type. Using the powerful psychology of Gurdjieff and Ouspensky, explore the automatic thoughts, attitudes and motives of your type, and discover the dynamics behind your relationships and the people you know. 168 pages. Hardcover, $19.95. Softcover, $9.95.

Written in such an easily read style you will wish it were longer. Recommended—The Unicorn

Maurice Nicoll, A Portrait

by Beryl Pogson. An account of Maurice Nicoll's life as a student of Gurdjieff and Ouspensky, and his teaching of the Fourth Way. Mrs. Pogson, Nicoll's secretary for fourteen years, describes Nicoll's personal life and methods of teaching, tells how he came to write the *Psychological Commentaries on the Teachings of Gurdjieff and Ouspensky*, establishes the course of his relationships with Gurdjieff and Ouspensky, and sheds light on his religious studies. 288 pages, 19 photographs. Softcover, $12.95.

A biography by a devoted pupil and private secretary of Nicoll, rich in personal detail and reminiscences. The most detailed published account of Nicoll's life as a student and teacher of the Gurdjieff work.—J. Walter Driscoll

Gurdjieff's Fourth Way: An Introduction

by Joel Friedlander. Verbatim transcripts of a series of seven lectures given at the New York Open Center designed to introduce the major ideas of the Gurdjieff system. Topics inclue states of consciousness, negative emotions, self-remembering and self-observation, cosmology, seven centers, and more. *Note: These lectures are available only by mail, and are not sold in any store.* 183 pages, illustrated. $65.00.

New Horizons; Explorations in Science

by P.D. Ouspensky, with a new Introduction by Colin Wilson. 216 pages. Softcover, $14.95.

Order Form / Mailing List Request

How to order:

Payment must accompany order. Please remember to calculate shipping charges according to the Shipping chart. For shipment to addresses in New York State, please add appropriate sales tax. Mail the completed order form (or a copy) with your payment to Globe Press Books at the address below. *Foreign Orders:* Surface shipping takes 2-15 weeks. Checks must be American Express or international checks drawn on a U.S. Bank. *Mailing List:* Simply fill in the name and address portion of the order form and return it to the address below.

Qty.	Title	Price	Total

Subtotal	
Tax	
Shipping	
Total	

Shipping		
	Surface	**Air**
U.S.A.	$1.75 first item, .50 each addtl.	$3.50 first item, 1.00 each addtl.
Canada & Mexico	$3.00 first item, 1.00 each addtl.	$4.50 first item, 2.00 each addtl.
Europe	$3.00 first item, 1.00 each addtl.	$8.50 first item, 3.00 each addtl.
Southern Hemisphere	$3.00 first item, 1.00 each addtl.	$10.00 first item, 5.75 each addtl.

> ### *Guarantee*
> Return any book in saleable condition within 30 days for a prompt and friendly refund.

Name _____

Address _____

_____

City / State / Zip _____

Country / Postcode _____

Mail to: Globe Press Books, P.O. Box 2045-N, Madison Square Station, New York, NY 10159.   *Thank you for your order.*